Chemical Safety in the Laboratory

Stephen K. Hall

LEWIS PUBLISHERS
Boca Raton   Ann Arbor   London   Tokyo

Library of Congress Cataloging-in-Publication Data

Hall, Stephen K.
    Chemical safety in the laboratory / Stephen K. Hall.
       p.   cm.
    Includes bibliographical references and index.
    ISBN 0-87371-896-8
    1. Chemical engineering laboratories—Safety measures.   I. Title.
TP187.H35   1994
660′.2804—dc20   93-28653
                                                          CIP

Preface

In the first quarter of this century, laboratories were normally found only in academic institutions, chemical industries, and hospitals. Now they are commonplace. Few industrial locations are without laboratory facilities of a size compatible with the range of operations and the substances being handled or used. Rapid advances and changes in science in general and chemistry in particular call for ever larger, more sophisticated, and better equipped laboratories. As the scope, size, and equipment of industrial laboratories become larger and more sophisticated, it must be paralleled in educational institutions in order that the skills of emergent laboratory workers keep pace with progress. Laboratories today operate on a scale unthought of a few decades ago.

This upsurge in laboratory activities has raised the level of consciousness of health and safety. The management effectiveness of any health and safety program is judged on the basis of how well it prevents injury and illness. By reducing accidents, injuries, and illnesses, organizations will be protecting their most important asset—workers. Nothing is more important to any organization than the safety and health of its workers. Thus, effective health and safety program management is vital to the continued success of every operation and organization.

While some organizations have excellent comprehensive laboratory safety and health programs, other organizations simply collect voluminous lists of hazardous substances, threshold limit values, compatible and incompatible chemicals and so on which give them a false sense of compliance with health and safety regulations. It is the apparent lack of a proper interpretation of such information that inspires the author to write this book. This book should be taken as a guide rather than a manual of procedures. Information contained in this book was obtained by the author through active involvement in managing and auditing safety and health programs over a period of 20 years. The aim is to provide a reference source for every person who works in, or uses the facilities of, laboratories.

The purpose of this book is to provide a proven approach to implementing and maintaining effective chemical hygiene programs in the laboratories. The book is designed for use by laboratory workers, as well as by supervisors and managers who are responsible for coordinating, directing, managing, and auditing safety and health programs; and chemical hygiene plans. The book may also be used by instructors in training and in courses on laboratory safety and health programs.

The author would like to thank Jon Lewis, Editor-in-Chief of Lewis Publishers, for his encouragement in preparing this book. His endorsement of the concept for this book was vital to its preparation. The author would also like to thank Cathleen Crawford for her invaluable editorial assistance throughout the preparation of the manuscript.

About the Author

Stephen K. Hall, Ph.D., has over 20 years of experience in the field of occupational safety and health. A corporate consultant on chemical safety, industrial hygiene and toxicology both in the U.S. and abroad, he has made a career of successfully achieving for his clients compliance with federal and state occupational safety and health regulations while addressing the topics of hazardous materials/waste management, emergency response and community right-to-know.

In addition to holding professional qualification as a registered environmental manager (REM), Dr. Hall is a certified hazardous materials manager (CHMM), a certified hazard control manager (CHCM), a certified hazardous waste specialist (CHWS), a registered medical technologist MT(ASCP), a certified clinical laboratory scientist (CLS), a registered sanitarian (RS), and a certified industrial hygienist (CIH) in toxicological aspects.

Dr. Hall is an alumnus of the University of Toronto, the University of Pittsburgh, and Harvard University. It was at Harvard University that he obtained his training in environmental health and toxicology. Dr. Hall has authored numerous journal articles, book chapters, and books on the subjects of chemistry, chemical safety, toxicology, occupational health, risk assessment, and hazardous materials/waste management. Due to his wide range of specialties and fluency in several languages, he is a noted keynote and seminar speaker at international conferences.

Disclaimer

The author of this book has been involved in laboratory health and safety activities for more than 20 years. The sources used to supplement the author's own practical experience are believed to be reliable and to represent the best opinion at the time of publication. The book is intended as a guide rather than a manual of procedures. No warranty, guarantee, or representation is made by the author nor by the publisher as to the correctness or adequacy or any information contained in the book. No responsibility or liability is accepted in any manner whatsoever for any errors or omissions.

Chemical Safety in the Laboratory

Contents

Chapter 1   The Occupational Safety and Health Act 1
Objectives of the OSHAct 2
OSHA Inspections 3
Violations and Penalties 4
OSHA Standards 5
State Programs 6

Chapter 2   The OSHA Laboratory Standard 9
Exemptions from the OSHA Laboratory Standard 10

Chapter 3   Employer's Responsibilities 13
General Responsibilities 14
Chemical Hygiene Responsibilities 15
Additional Responsibilities 16

Chapter 4   The Chemical Hygiene Plan 19
Standard Operating Procedures 19
Chemical Hazard Control Measures 19
Laboratory Ventilation 20
Employee Information and Training 20
Prior Approval 21
Medical Consultation and Examination 21
Designation of a Chemical Hygiene Officer 22
Special Procedures 23

Chapter 5   General Safety Practices 25
General Safety Rules 26
Personal Hygiene and Safety Rules 27
Laboratory Glassware Safety Rules 27
Housekeeping Safety Rules 28
Miscellaneous Safety Rules 28

Chapter 6   Personal Protective Equipment 31
Eye and Face Protection 32
Body Protection 34
Hand Protection 35
Respiratory Protection 36
Standard Safety and Emergency Equipment 44

Chapter 7   Identification of Hazardous Chemicals 47
The United Nations Numbering System 47
The U.S. Department of Transportation System 48

The DOT Placarding/Labeling System 54
The NFPA Hazard Identification System 54

Chapter 8   Chemical Handling Procedures 59
Handling Procedures for Flammable Chemicals 60
Handling Procedures for Pyrophoric Chemicals 64
Handling Procedures for Hypergolic Mixtures 65
Handling Procedures for Reactive/Unstable Chemicals 66
Handling Procedures for Water-Reactive Chemicals 70
Handling Procedures for Corrosive Chemicals 71
Handling Procedures for Toxic Chemicals 74

Chapter 9   Storage of Chemicals 81
Inventory Control 81
Storage According to Chemical Compatibility 82
Procedures for the Storeroom/Stockroom 84
Storage of Flammable Chemicals 86
Storage of Oxidizers 87
Storage of Water-Reactive Chemicals 87
Storage of Corrosive Chemicals 88
Storage of Toxic Chemicals 88

Chapter 10   Laboratory Ventilation 91
Dilution vs. Local Exhaust 91
OSHA Ventilation Standards 92
Local Exhaust Systems 92
Types of Hoods 93
Testing of Hoods 97
Duct Velocity Measurement 99
Laboratory Fume Hoods 99

Chapter 11   Employee Chemical Exposure Monitoring 105
Air Contaminants in the Laboratory 106
Environmental Factors Affecting Monitoring 106
General Air Monitoring Criteria 107
Air Monitoring Considerations 107
Monitoring Procedures and Devices 110
Analysis of Air Samples 115
Interpretation of Results 115

Chapter 12   Medical Examination and Consultation 117
Medical Program Development 117
Pre-Employment Medical Examination 118
Periodic Medical Examination and Consultation 120

Emergency Medical Treatment 121
Non-Emergency Medical Treatment 122
Medical Record Keeping 123
Medical Program Review 123
Medical Examination upon Termination 124

Chapter 13   Chemical Emergency Planning 125
Chemical Emergency Officer 125
Potential Chemical Emergencies 126
Magnitude of Injury or Damage 127
Chain of Hazard Evaluation 127
Prevention Method 127
Chemical Emergency Response Procedure 127
Site Map 128
Communication 128
Alarm System 128
Control of Utility Services 129
Site Security and Control 129
Evacuation 129
Chemical Emergency Equipment 130
Rescue 130
Decontamination 131
Emergency Medical Treatment 131
Training for Chemical Emergencies 131
Follow-Up and Review 132

Chapter 14   Chemical Emergency Response Procedure 133
Personnel Protection 134
Hazard Area Control 135
Chemical Hazard Exposure Assessment 137
Levels of Exposure 140
Role of the First Responder 141
Initial Response Action 142
Containment Techniques 145
Leak-Stopping Devices 147
Emergency Response Procedures for Flammable Chemicals 149
Emergency Response Procedures for Oxidizers 151
Emergency Response Procedures for Organic Peroxides 152
Emergency Response Procedures for Toxic Chemicals 152
Emergency Response Procedures for Corrosive Chemicals 153

Chapter 15   Decontamination of Emergency Responders 155
Minimizing Contamination 155
Decontamination Plan 156

Decontamination Methods 157
Testing for Effectiveness 158
Sample Decontamination Procedure 158

Chapter 16   Victim Handling Procedures 165
Stages of Emergency Response 165
Assessment of Victims 166
Triage 167
Initial Stabilization 169
Decontamination of Victims 171
Evacuation of Victims 172
Heat Stress Victims 172
Administration of First Aid 175

Chapter 17   Laboratory Waste Management 179
Definition of Hazardous Waste 180
Categories of Generators 184
On-Site Waste Storage 185
Hazardous Waste Shipment Preparation 186
The Uniform Hazardous Waste Manifest 187
Waste Reduction 188

Chapter 18   Employee Training 189
Training Program Development 189
The Trainer 190
Training for Supervisors 191
New Employee Training 192
Training for Transferred Employees 194
Keeping Record of Training Activities 194

Chapter 19   Record Keeping 195
Purpose of Section 1910.20 195
Preservation of Records 196
Access to Records 196
Trade Secrets 197
Employee Information 197
Transfer of Records 198

Bibliography 199

Appendix A. States with OSHA-Approved Plans 203
Appendix B. The OSHA Laboratory Standard 207

Appendix C.   EPA Levels of Protection 217
Appendix D.   Vapor Suppression Foams 221
Appendix E.   EPA-Listed RCRA Hazardous Wastes 223
Appendix F.   Incompatible Chemicals in Storage and in Reactions 249

Index 251

List of Tables

Table 6-1.   OSHA Standards on Personnel Protection 32
Table 6-2.   Chemical Resistance of Common Glove Materials 37
Table 7-1.   United Nations Hazard Classes 48
Table 7-2.   Some Hazardous Materials by Class and Identification
             Number 49
Table 7-3.   DOT Classification of Hazardous Materials 50
Table 7-4.   DOT Placarding System 55
Table 8-1.   NFPA Classification of Flammable and Combustible
             Liquids 61
Table 8-2.   Properties of Some Common Flammable Liquids 62
Table 8-3.   Some Common Pyrophoric Chemicals 65
Table 8-4.   Some Common Reactive Compounds 68
Table 9-1.   Dangers of Storing Chemicals in Alphabetical Order 83
Table 10-1. Recommended Face Velocity of Laboratory Fume
             Hoods 103
Table 12-1. Recommended Medical Surveillance Program 119
Table 12-2. Medical Tests 121
Table 14-1. Commonly Used Absorbents 146
Table 16-1. Glasgow Coma Scale 170
Table 17-1. Criteria for TCLP 183

List of Figures

Figure 6-1   Flow chart for respiratory selection 41
Figure 7-1   NFPA diamond symbol 56
Figure 8-1   Flammability factors of a flammable liquid 63
Figure 10-1   Enclosure type of hood 94
Figure 10-2   Receiving type of hood 95
Figure 10-3   Capturing type of hood 96
Figure 10-4   Conventional fume hood 100
Figure 10-5   Bypass-air fume hood 101
Figure 10-6   Add-air fume hood 102
Figure 11-1   Evacuated flasks 111
Figure 11-2   Gas or liquid displacement collector 112
Figure 11-3   Absorbers 113
Figure 11-4   Activated charcoal tube 114
Figure 14-1   Hazard area control 137
Figure 15-1   Sample decontamination stations and layout 160
Figure 17-1   A "Lab Pack" 187

1   THE OCCUPATIONAL SAFETY AND HEALTH ACT

In December 1970, Congress passed and President Richard M. Nixon signed into law the Williams-Steiger Occupational Safety and Health Act (OSHAct) of 1970, which became effective on April 28, 1971. The Act applies to all businesses involved in or affecting interstate commerce. Exempt are employees already covered by another federal safety program, such as those under the Atomic Energy Act of 1954 and the Coal Mine Health and Safety Act of 1969. In addition, government agencies such as the Department of Defense have directed that the standards established by the OSHAct will be observed.

The OSHAct set up several new organizations. One is the Occupational Safety and Health Administration (OSHA), a new organization within the Department of Labor, to set safety and health standards for almost all nongovernmental employers. The Department of Labor is responsible for issuing and enforcing occupational safety and health standards. To ensure that the standards are observed, inspections are made by OSHA compliance officers located in offices throughout the country.

Another new organization is the Occupational Safety and Health Review Commission (OSHRC), a quasijudicial board composed of three members appointed by the President. Functions of the OSHRC are to hear and review alleged violations and, where warranted, to issue corrective orders and to assess civil penalties.

A third new organization is the National Institute for Occupational Safety and Health (NIOSH), within the Department of Health and Human Services. NIOSH's function is to carry out research and educational functions. It is authorized to conduct research on hazardous substances and conditions to permit tolerance levels to be established as parts of the OSHA standards. It develops and recommends to the Secretary of Labor new occupational safety and health standards. To assist in developing these recommendations, NIOSH may issue regulations requiring employers to measure, record, and report on employees' exposures to substances or physical agents that endanger health and safety. NIOSH publishes annually lists of known toxic substances and the levels at which toxic effects will occur. At the request of an employer or employees' representative, NIOSH will determine whether any substance found in the place of employment has potential hazardous effects. In addition, NIOSH will conduct and publish studies on industrial processes, substances, and stresses for their potential effects on personnel.

OBJECTIVES OF THE OSHAct

The fundamental objective of the OSHAct is to ensure "so far as possible every working man and woman in the nation safe and healthful working conditions and to preserve our human resources." This fundamental objective can be achieved by:

(1) Encouraging employers and employees in their efforts to reduce the number of occupational safety and health hazards at their places of employment, and to stimulate employers and employees to institute new and to perfect existing programs for providing safe and healthful working conditions;

(2) Providing that employers and employees have separate but dependent responsibilities and rights with respect to achieving safe and healthful working conditions;

(3) Authorizing the Secretary of Labor to set mandatory occupational safety and health standards applicable to businesses affecting interstate commerce, and by creating an Occupational Safety and Health Review Commission for carrying out adjudicatory functions under the Act;

(4) Building upon advances already made through employer and employee initiative for providing safe and healthful working conditions;

(5) Providing for research in the field of occupational safety and health, including the psychological factors involved, and by developing innovative methods, techniques, and approaches for dealing with occupational safety and health problems;

(6) Exploring ways to discover latent diseases, establishing causal connections between diseases and work in environmental conditions, and conducting other research relating to health problems, in recognition of the fact that occupational health standards present problems often different from those involved in occupational safety;

(7) Providing medical criteria which will assure, insofar as practicable, that no employee will suffer diminished health, functional capacity, or life expectancy as a result of his work experience;

(8) Providing for training programs to increase the number and competence of personnel engaged in the field of occupational safety and health;

(9) Providing for the development and promulgation of occupational safety and health standards;

(10) Providing an effective enforcement program which shall include a prohibition against giving advance notice of any inspection and sanctions for any individual violating this prohibition;

(11) Encouraging the states to assume the fullest responsibilities for the administration and enforcement of their occupational safety and health laws by providing grants to the states to assist in identifying their needs and responsibilities in the area of occupational safety and health, to develop plans in accordance with the provisions of the OSHAct, to improve the administration and enforcement of state occupational safety and health laws, and to conduct experimental and demonstration projects in connection therewith;

(12) Providing for appropriate reporting procedures with respect to occupational safety and health, which procedures will help achieve the objectives of the OSHAct and accurately describe the nature of the occupational safety and health problems; and

(13) Encouraging joint labor-management efforts to reduce injuries and disease arising out of employment.

OSHA INSPECTIONS

To ensure compliance with the OSHA standards, inspections are made without prior notice by OSHA compliance officers. Compliance officers have the right to enter the premises of any business, to take air, water, or other environmental samples; interview employees; and inspect all records related to safety and health. In general, the following priorities are used in making inspections:

(1) An inspection will follow any accidental death or mishap in which five or more workers are injured. Under the law, such an accident must be reported within 48 hr of its occurrence.

(2) A plant will be inspected if a report is received of an imminent hazard, and to ascertain that the imminent hazard reported has been eliminated.

(3) Industries that are considered especially hazardous will be inspected at frequencies and times established by the cognizant OSHA office. The especially hazardous production industries are those with high injury frequency rates, e.g., longshoring, lumber, wood, roofing, sheet metal, meat, meat products, mobile homes and various transportation equipment.

(4)   Other industries, also on schedules to be established by OSHA offices.

VIOLATIONS AND PENALTIES

If the OSHA compliance officer finds a violation of the OSHA standard, he or she must issue a citation to the employer within 6 months of any violation. The citation will indicate a reasonable time for elimination or abatement of the hazard. Four types of citations for violations of OSHA standards are possible:

(1)   *Imminent Danger:* any condition or practice such that a danger exists which could reasonably be expected to cause death or serious physical harm immediately or before correction can be made through normal procedures.

(2)   *Serious Violation:* any condition, practice means, methods, operation, or process which has a substantial probability of causing deaths or serious physical harm.

(3)   *Nonserious Violation:* where an incident or occupational illness resulting from violation of a standard would probably not cause death or serious physical harm (no permanent injury).

(4)   *De Minimis:* (no penalty): where violation of standard has no immediate or direct relationship to safety or health.

Plant housekeeping is one area graded by the OSHA compliance officer. In addition to constituting a hazard that could contribute to falls, fires, and other accidents, housekeeping is indicative of the general management of the plant, especially in its attitude toward safety. Poor housekeeping will be immediately apparent to any compliance officer entering a plant and will undoubtedly influence his or her attitude toward the plant manager's concerns for employee health and safety.

The compliance officer who finds conditions of imminent danger can only request, not demand, shutdown of an operation. If the employer or his representative refuses to shut down the operation, the compliance officer can, and will, notify the employees of the imminent hazard, and the Department of Labor may seek court authority to shut down the operation.

For any violation, the inspecting compliance officer can propose a penalty of up to $1000 per day until the situation is corrected. For serious violations, such penalties are mandatory. Such penalties may also be imposed where nonserious violations are cited.

Willful or repeated violations may each be assessed a civil penalty of up to $10,000. If the death of an employee is due to a willful violation, the employer, if convicted in court, may be punished by a fine of not more than $70,000, imprisonment of not more than 6 months, or both. If death of an employee occurs after the employer has been once convicted of a violation, the penalty can be doubled.

At the conclusion of an inspection, the compliance officer conducts an interview with management and presents his or her findings. The compliance officer then reports the findings to the Director, who decides whether a citation, with an abatement date, is to be issued, and the penalty to be assessed. A citation is sent by registered mail to the employer.

Any employer cited for violation can contest any citation, proposed penalty, or time stipulated to eliminate or abate a hazard. A judge from the OSHRC will review the case, decide whether a violation has occurred, and if so, the penalty. The judge's decision may be reviewed by the three-member Commission, and it may be appealed to the U.S. Appeals Court. Employees or their representatives may also object, within 15 days after a citation is issued, to the time stipulated for eliminating or abating a violation.

Where elimination or abatement of a cited hazard cannot be accomplished within the allotted time, an extension may be granted if the employer shows that he has made a good faith effort to do so, but that factors beyond the employer's control are causing the delay.

OSHA STANDARDS

The OSHAct empowers the Secretary of Labor to set safety and health standards. Until April 1973 the Secretary could stipulate unilaterally that any recognized consensus standard, produced by an existing organization that undertakes such tasks, would be observed. There was no need for any formal hearings to determine whether such standards should be issued. After that date hearings must be held to include new or revised standards, those that are not recognized consensus standards, or those that were not already existing federal standards.

The first version of the OSHA standards was a hastily prepared assembly of standards previously issued by technical associations to meet the April 1973 cutoff date. Consensus standards included those by the American Conference of Governmental Industrial Hygienists (ACGIH), American National Standards Institute (ANSI), National Fire Protection Association (NFPA), American Society of Mechanical Engineers (ASME), and others. These standards had been prepared as recommended guidelines to be observed as deemed advisable by industry. In certain instances they became legally mandatory when a political jurisdiction, such as a state or city, requires observance of such provisions, e.g., the ASME Code for Pressure Vessels. Many of these also became economically mandatory where insurance companies raised rates prohibitively due to nonob- servance. Other standards, such as those previously issued under the Walsh- Healey Act, had already become mandatory. The Walsh-Healey Act require- ments were imposed on any company which received a federal contract valued at $10,000 or more. The modifying of the OSHA standards is expected to be a continuing process. Changes will occur as experience indicates need, new criteria will be adopted, old ones updated to conform with new practices and processes, and unneeded ones eliminated.

In recent years, OSHA has eliminated numerous "nitpicking" rules that were considered more of a nuisance than a help in protecting workers. This evolution has reduced criticism directed toward the agency since the standards were first issued. According to OSHA, this lessening of criticism permits the agency to concentrate on reducing or eliminating the more serious and significant workplace safety and health hazards. In addition, as the compliance officers gain experience in recognizing which hazards and safeguards are significant, it is expected that adverse comments on their performance will gradually diminish.

The OSHA standards consists of the following four categories:

(1) *Design Standards:* fixed requirements that stipulate dimensions, materials, types of warnings, and so forth. Examples of these design standards are ventilation design details contained in Section 1910.94 of the initial standards.

(2) *Performance Standards:* indicate the end result to be achieved, but not the methodology by which it is to be accomplished. Examples of performance standards are the Threshold Limit Values (TLVs) of the ACGIH, which are contained in Section 1910.1000.

(3) *Horizontal-type Standards:* the provisions are enough to apply to a wide variety of operations or types of employment. Examples of horizontal standards are sanitation (1910.141) and walking and working surfaces (1910 Subpart D).

(4) *Vertical-type Standards:* the provisions of each of these standards are developed for and apply only to one type of employment. Construction, ship building, ship repairing, ship breaking, and longshoring were the first to have vertical-type standards.

The OSHA standards require the accumulation of numerous records by each employer. The need for preparation of these records and the costs entailed were the major factors that raised the most strenuous objection of the OSHA standards. Most of these are maintained at the place of employment, where they will be available for inspection and use by the OSHA compliance officer during any visit he or she might make there. There are penalties for failure to comply with the record-keeping requirements in any OSHA standard.

STATE PROGRAMS

One of the reasons for the passage of the OSHAct was the inadequacies of existing state programs. In many instances, state safety codes and regulations were lacking in their provisions and very badly enforced. The lack was the result of the scantiness of funding and the quality of the personnel empowered to enforce the laws. If the specific conditions are met, the OSHAct permits the states to regain sole authority to police occupational safety.

A state wanting to regain control over industries within its borders covered by the OSHAct must submit a proposed plan indicating how it intends to carry out a program which at the minimum is as effective as the federal program. The state plan must be approved by the Secretary of Labor. The state receiving approval will then pay half the cost of the program, the remainder of the funding is paid by the federal government. The Department of Labor will maintain surveillance over the state program for 3 years to ensure that the state carries out its responsibilities. Currently, a total of 25 states have approved programs. (Appendix A).

2 THE OSHA LABORATORY STANDARD

Many academic, industrial, and other laboratories that use hazardous chemicals are now required by federal regulations known as the "OSHA Laboratory Standard" (29 CFR 1910.1450) to develop and implement chemical hygiene plans. (Appendix B). A *hazardous chemical* is defined in the OSHA Laboratory Standard as any chemical substance for which there is statistically significant evidence based on at least one study conducted in accordance with established scientific principles that acute or chronic health effects may occur in exposed workers. The term *health hazard* includes chemicals that are carcinogens, toxic or highly toxic agents, reproductive toxins, irritants, corrosives, sensitizers, hepatotoxins, nephrotoxins, neurotoxins, agents that act on the hematopoietic systems, as well as agents that damage the lungs, skin, eyes, or mucous membrane.

The OSHA Laboratory Standard is a *performance standard,* i.e., there are few specific requirements to carry out certain procedures in a certain way. Specific results to be achieved are particularly denoted, but the manner by which the results are to be accomplished is not delineated. It differs from many OSHA health standards in that it does not establish new exposure limits but sets other performance provisions designed to protect laboratory workers from potential hazards in their work environment. To the extent possible, it allows a large measure of flexibility in compliance measures; special provisions must be explicitly considered by the employer, but need only be implemented when the employer deems them necessary on the basis of specific conditions existing in the laboratory. The primary emphasis of the standard is on administrative controls necessary to protect workers from overexposure to hazardous chemicals in laboratories.

In addition to protecting laboratory workers from potential harm due to exposure to chemicals, for the purposes of the OSHA Laboratory Standard, the

9

term *laboratory workers* is extended to include office, custodial, maintenance, and repair personnel, as well as others who, as part of their duties, regularly spend a significant amount of their working time within the laboratory environment. It is the responsibility of an employer to determine what constitutes a "significant amount" of working time. An employer's determination is, of course, subject to review at the time of visit by an OSHA compliance officer.

EXEMPTIONS FROM THE OSHA LABORATORY STANDARD

The OSHA Laboratory Standard does not apply to all laboratories. Where it does apply, it supersedes the requirements of all other OSHA health standards in 29 CFR 1910, subpart Z. For example, even though some laboratories may already have implemented the Hazard Communication Standard (29 CFR 1910.1200), the OSHA Laboratory Standard takes precedence in those laboratories to which it applies. Employers do not have the option of choosing between the two standards. If the OSHA Laboratory Standard applies, the employer must have it implemented. If the OSHA Laboratory Standard does not apply, then the Hazard Communication Standard, or other health standards, will apply.

The OSHA Laboratory Standard does not apply to uses of hazardous chemicals that do not meet the definition of laboratory use, and in such cases, the employer shall comply with the relevant OSHA standard in 29 CFR 1910, even if such use occurs in a laboratory. For example, chemical plant quality control laboratories might be exempt because they are usually adjuncts of production operations that typically perform repetitive procedures for the purpose of a product, or process. Likewise, pilot plant operations might also be exempt. However, if the pilot plant operation is an integral part of a research function for the purpose of evaluating a particular effect, then that pilot plant operation may be covered under the OSHA Laboratory Standard.

Laboratories that may use hazardous chemicals but provide no potential for employee exposure might be exempt. An example of such condition might include procedures using chemically impregnated test media such as dip-and-read tests where a reagent strip is dipped into the specimen to be tested and the results are interpreted by comparing the color reaction to a color chart supplied by the manufacturer of the test strip. Another example might include certain commercially prepared kits such as those used in performing pregnancy tests in which all of the reagents needed to conduct the test are contained in the kit.

Some laboratories that may come under requirements of a substance-specific standard might also be exempt. For example, laboratories involved in histology, pathology, and human or animal anatomy must comply with the specific requirements of the Formaldehyde Standard. Currently, substance-specific standards have been established for the following chemicals: 2-acetylaminofluorene, acrylonitrile, 4-aminodiphenyl, arsenic and its inorganic compounds, asbestos (actinolite, anthophyllite, tremolite), benzene, benzidine, bis(chloromethyl)ether, coal tar pitch volatiles, coke oven emissions, cotton dust, 1,2-dibromo-3-chloropropane, 3,3′-dichlorobenzidine (and its salts), 4-dimethylaminoazobenzene, ethylene oxide, ethyleneimine, formaldehyde, lead, methyl chloromethyl ether, α-naphthylamine, β-naphthylamine, 4-nitrobiphenyl, N-nitrosodimethylamine, β-propiolactone, vinyl chloride.

3

EMPLOYER'S RESPONSIBILITIES

Both employers and employees have responsibilities they must carry out under the act, although employers are the only ones who can be penalized for failure to comply with the OSHAct. The fundamental requirement on the employer is similar to the employer's responsibility under the common law: to "furnish to each of his employees employment and a place of employment free from recognized hazards that are causing or are likely to cause death or serious physical harm to his employees." In addition, the employer must comply with the OSHA standards for the industry; keep records of work-related injuries, illnesses, and deaths; and keep records of exposure of employees to toxic materials and harmful physical agents. The employer must also notify employees of the provisions of the law, their protections and obligations. The employer must keep employees informed on matters of safety and health and on accidents and alleged safety violations in the place of employment. The employer must refrain from discriminating against employees who file complaints regarding hazardous working conditions.

The employee is obligated to "comply with occupational safety and health standards and all rules, regulations, and orders issued pursuant to the OSHAct that are applicable to his or her own actions and standards." In addition, the employee or the employee's representative may file complaints of violations with the Department of Labor or with the compliance officer who may inspect the establishment. An employee may accompany the compliance officer during an inspection. Further, employees may submit recommendations on new standards to be promulgated.

If an employee fails to observe the prescribed safety and health standards, the employer may be cited for a violation. For example, a chemical company was cited and fined for violation by workers that included: not wearing protective goggles or shields, poor housekeeping, oxygen cylinders lying near acetylene-filled cylinders, gas cylinders whose valves were not closed while not in use, and improper electrical equipment grounding.

13

GENERAL RESPONSIBILITIES

All laboratories that fall under the OSHA Laboratory Standard have, as employers, certain obligations. Some of these obligations are summarized below:

Minimize All Chemical Exposures

Few laboratory chemicals are without hazards. Because of this fact, general precautions for handling all laboratory chemicals should be adopted, as opposed to specific guidelines for particular chemicals. As a cardinal rule, skin contact with all chemicals should be avoided.

Avoid Underestimation of Risk

Even for chemicals without known significant hazard, all exposure should be minimized. Needless to say, in working with chemicals with known hazards special precautions must be taken. It is prudent to assume that any mixture will be at least as toxic as its most toxic component, and all chemicals of unknown toxicity are potentially toxic.

Provide Adequate Ventilation

The best way to prevent worker exposure to airborne chemicals is to prevent their escape into the workplace atmosphere through the use of ventilation, or other engineering controls. Ventilation should be the method of choice to dilute contaminants to safe levels or to capture and remove them at their sources of generation.

Observe the PELs and TLVs

The permissible exposure limits (PELs) of OSHA and the threshold limit values (TLVs) of the American Conference of Governmental Industrial Hygienists (ACGIH) are not to be routinely exceeded. TLVs are time-weighted average concentrations of chemical contaminants under which most workers can work consistently for 8 hr a day, day after day, with no harmful effects. PELs are published and enforced by OSHA as legal standards.

Institute a Chemical Hygiene Plan

A mandatory chemical hygiene program designed to minimize laboratory worker exposure to chemicals must be developed and implemented. Its implementation must be observed by the employer as well as by all laboratory workers. A sound chemical hygiene program sets forth the work practices and procedures, as well as personal protective equipment and other equipment that will protect laboratory workers from harm arising from hazardous chemicals used in the laboratory.

CHEMICAL HYGIENE RESPONSIBILITIES

All personnel associated with laboratory operations are to share chemical hygiene responsibilities. Some of these responsibilities are summarized below:

Chief Executive Officer

While the development and implementation of a chemical hygiene plan is a responsibility that rests at all levels, the chief executive officer has ultimate responsibility within the organization and must provide continuing support for the program.

Supervisor of Administrative Unit

The supervisor of an administrative unit acts as the representative of the chief executive officer in the capacity of implementing a chemical hygiene plan and must vigorously oversee its implementation.

Chemical Hygiene Officer

The employer appoints this individual to develop and implement a chemical hygiene plan. In order to provide proper guidance, this individual should be qualified by training and/or experience. This assignment can be a second title for a person who has other responsibilities in the organization. In any organization, there may well be more than one person serving as a chemical hygiene officer because the organization may have several distinctly different laboratories. Thus, a director or supervisor of a laboratory or a principal investigator of a research grant might each be a suitable person to develop and implement a chemical hygiene plan for their respective laboratory and to be designated as a chemical hygiene officer by the chief executive officer.

Laboratory Supervisor

The responsibility of this individual is chemical hygiene in the specific laboratory he supervises. His responsibilities include, but are not limited to, the insurance that workers in his laboratory know and follow the chemical hygiene plan rules, that protective equipment is available and in working order, and that appropriate training has been provided. The laboratory supervisor should also know the current legal requirements concerning regulated chemical substances, determine the required levels of personnel protection, ensure the adequacy of training for use of any equipment and material, and provide regular, formal chemical hygiene and housekeeping inspections, as well as routine inspections of emergency equipment.

Project Director

The primary responsibility of this individual is the monitoring of chemical hygiene procedures for that specific laboratory operation.

Laboratory Worker

All laboratory workers have the responsibility of performing each laboratory operation in accordance with the chemical hygiene procedures of the organization and to develop safe personal chemical hygiene habits.

ADDITIONAL RESPONSIBILITIES

In addition to the general and chemical hygiene responsibilities, the employer also has additional responsibilities and these are summarized below:

Employee Exposure Monitoring Program

The employer conducts air monitoring to determine the laboratory worker's exposure to any hazardous chemical regulated by an OSHA standard requiring monitoring if there is reason to believe that exposure levels for that chemical routinely exceed the action level, or in the absence of an action level, either the

PEL, or the TLV. If the initial monitoring discloses exposure above the action level, the employer should immediately comply with the exposure monitoring provisions of the relevant OSHA standard.

The employer, within 15 working days after the receipt of any air monitoring results, is to notify the laboratory worker of such results in writing either individually or by posting results in an appropriate location that is accessible to all laboratory workers. Laboratory exposure monitoring may be terminated in accordance with the relevant OSHA standard.

Employers should recognize that laboratory worker exposure monitoring is not a trivial procedure and it must be performed by a qualified industrial hygienist, or other competent professional.

Hazard Determination Program

The employer should apply the following provisions in the determination of potential hazards in the laboratory:

(1) If the composition of the chemical substance that is produced exclusively for the laboratory's use is known, the employer should determine if it is a hazardous chemical as defined in the OSHA Laboratory Standard. If the chemical substance is determined to be hazardous, the employer should provide appropriate employee information and training as outlined in the chemical hygiene plan.

(2) If the chemical substance produced is a byproduct of unknown composition, the employer should assume that the chemical substance is toxic and hazardous and implement the chemical hygiene plan.

(3) If the chemical substance is produced for another user outside of the laboratory, the employer should comply with the Hazard Communication Standard (29 CFR 1910.1200) including the requirements for preparation of MSDSs and labeling.

Chemical Emergency Response Plan

Chemicals spills and accidents are frequent occurrences in the laboratory. The employer must develop a chemical emergency policy that includes consideration of prevention, containment, clean-up, decontamination, and reporting. A written emergency response plan must be established and communicated to all laboratory employees. The plan must include procedures for ventilation failure, evacuation, decontamination, medical care, reporting, and drills. All chemical accidents, or anytime the emergency response plan has been set in motion, the incident should be carefully analyzed with the results distributed to all who might benefit.

Laboratory Waste Management Program

The employer is to develop a comprehensive laboratory hazardous waste program to specify how laboratory wastes are to be segregated, stored, collected, and transported in accordance with current regulations of the Environmental Protection Agency (EPA) as well as the Department of Transportation (DOT).

Respirator Program

Laboratory workers are to wear respirators whenever they might be exposed to airborne concentrations of hazardous chemicals greater than the action level, either the PEL, or the TLV. Where the use of a respirator is necessary, the employer must provide the proper equipment at no cost to the laboratory worker and all respiratory equipment must be selected and used in accordance with the OSHA Respirator Standard (29 CFR 1910.134), including in particular the following:

(1)   A written respirator program governing the selection and use of respirators; and

(2)   All laboratory workers who are likely to need to use respirators must be trained in their proper use, inspection, and maintenance.

Record Keeping

The employer has the responsibility to establish and maintain for each laboratory worker an accurate record of any environmental measurements taken to monitor worker exposures, and any medical consultation and examinations, including tests or written opinions, required by the OSHA Laboratory Standard. The employer should also assure that such records are kept, transferred, and made available in accordance with the requirements of 29 CFR 1910.20.

4   THE CHEMICAL HYGIENE PLAN

A chemical hygiene plan is a written document developed and implemented by the employer that sets forth procedures, equipment, personal protective equipment, and work practices that are capable of protecting laboratory workers from the health hazards presented by the hazardous chemicals used in that particular laboratory and keeping exposures below applicable OSHA exposure limits. The chemical hygiene plan should be made readily available to employees, their designated representatives and to the Assistant Secretary of Labor upon request. The employer should review and evaluate the effectiveness of the chemical hygiene plan at least annually and have it updated as necessary.

A chemical hygiene plan should include each of the following elements and should indicate specific measures that the employer will take to ensure protection of its laboratory workers.

STANDARD OPERATING PROCEDURES

Standard operating procedures relevant to safety and health considerations are to be established and followed by all laboratory workers when work involves the use of hazardous chemicals. General standard operating procedures are to be established as well as hazard-specific safety procedures.

CHEMICAL HAZARD CONTROL MEASURES

Criteria that the employer will use to determine and implement control measures to reduce worker exposure to hazardous chemicals including engineering controls, the use of personal protective equipment, and established personal hygiene practices. Particular attention should be given to the selection of control measures for chemicals that are known to be extremely hazardous.

Chemical safety is achieved by continual awareness of chemical hazards and by keeping the chemical, and its potential hazard, under control by proper use of control measures and equipment. Laboratory workers must be familiar with the precautions to be taken and be alert to the possible malfunction of any safeguard equipment. All engineering safeguards and control measures must be properly maintained and inspected on a regular basis, and are never to be overloaded beyond their design limits.

LABORATORY VENTILATION

A laboratory is to have an appropriate general ventilation system with air intakes and exhausts located far enough apart to avoid recirculation of contaminated air. This ventilation system is to provide a source of air for breathing and for input to local ventilation devices. Laboratory general ventilation should be between 4 and 12 room air changes per hr. In addition, local exhaust systems such as laboratory fume hoods or other ventilation devices should be used as the primary method of control to prevent accumulation of chemical vapors. General air flow should be relatively uniform throughout the laboratory, without any turbulence, high velocity, or static areas.

Laboratory fume hood face velocity should typically be between 75 and 150 linear ft per minute (fpm) for most chemicals. A face velocity of 150 fpm or higher might be desirable for highly toxic chemicals. Any alteration or modification of the ventilation system is to be made only after thorough testing indicates that worker protection from airborne toxic chemicals will continue to be adequate. Fume hoods and other chemical-hygiene-related equipment should undergo continuing appraisal and be modified if inadequate.

EMPLOYEE INFORMATION AND TRAINING

The employer is required under the OSHA Laboratory Standard to provide employees with information and training to ensure that they are apprised of the hazards of chemicals present in their work areas. Such information is provided at the time of an employee's initial assignment to a work area where hazardous chemicals are present and prior to assignment involving new exposure situations. The frequency of refresher information and training is to be determined by the employer.

Employees must be informed of and given access to the contents of the OSHA Laboratory Standard and its appendices, the employer's chemical hygiene plan, the PELs or TLVs of the hazardous chemicals used in the laboratory, the signs and symptoms associated with exposures to such chemicals, as well as the location and availability of known reference material on the hazards, safe handling, storage, and disposal of hazardous chemicals including the material safety data sheets (MSDSs).

Employee training should include methods and observations that may be used to detect the presence or release of a hazardous chemical, the physical and health hazards of chemicals in the laboratory, and the measures the employees can take to protect themselves from these hazards, including specific procedures the employer has implemented to protect employees from exposure to hazardous chemicals, such as appropriate work practices, emergency procedures, and personal protective equipment to be used.

In summary, all laboratory workers must be informed of and trained on the applicable aspects of the employer's written Chemical Hygiene Plan.

PRIOR APPROVAL

There are circumstances under which a particular laboratory operation, procedure, or activity may require prior approval from the employer, or the employer's representative, before implementation. Such laboratory operations, procedures, or activities must obtain a written description of specific safety practices incorporating the applicable precautions. Employees must understand these practices and obtain prior approval before commencing such an operation.

MEDICAL CONSULTATION AND EXAMINATION

The employer must provide all laboratory workers an opportunity to receive medical attention after an exposure to a hazard, including any follow-up examination that the examining physician determines to be necessary. For example, whenever a worker develops signs or symptoms associated with an exposure to a hazardous chemical to which the worker may have been exposed in the laboratory, the worker shall be provided an opportunity to receive an appropriate medical examination. Another example is that where exposure monitoring reveals an exposure level routinely above the action level, or in the absence of an action level, either the PEL, or the TLV, for an OSHA regulated substance for which there are exposure monitoring and medical surveillance requirements, medical surveillance is to be established for the affected worker as prescribed by that particular OSHA standard.

Whenever a spill, leak, explosion, or other chemical accident occurs that may result in the likelihood of an exposure to a hazardous chemical, the affected laboratory worker must be provided with an opportunity for a medical consultation. Such consultations are necessary for the purpose of determining the need for a medical examination.

All medical consultations and examinations are to be performed by or under the direct supervision of a licensed physician and provided without cost by the employer to the laboratory worker without loss of pay and at a reasonable time and place. It is the responsibility of the employer to provide the physician with the following information:

(1) The identity of the hazardous chemical the laboratory worker may have been exposed to;

(2) A description of the conditions under which the exposure occurred including quantitative data, if available; and

(3) A description of the signs and symptoms of exposure that the laboratory worker is experiencing.

For all medical consultation and examination required under the OSHA Laboratory Standard, the employer should obtain a written opinion from the attending physician. The written opinion should include the following information:

(1) Any recommendation for further medical follow-up;

(2) The results of the medical examination and associated tests;

(3) Any medical condition that may be revealed in the course of the examination that may place the laboratory worker at increased risk as a result of exposure to a hazardous chemical present in the laboratory; and

(4) A statement that the laboratory worker has been informed by the physician of the results of the consultation or medical examination and any medical condition that may require further examination or treatment.

The physician's written opinion must not reveal specific findings of diagnosis unrelated to occupational exposure to any laboratory chemical.

DESIGNATION OF A CHEMICAL HYGIENE OFFICER

The employer must designate an individual qualified by training or experience as the chemical hygiene officer of a laboratory or a group of laboratories to be responsible for the development and implementation of the chemical hygiene plan. If appropriate, a chemical hygiene committee must be established. The responsibilities of the chemical hygiene officer must be described in the chemical hygiene plan.

SPECIAL PROCEDURES

The employer must provide for additional employee protection for work with chemicals that have a high degree of acute toxicity and chemicals of unknown toxicity, e.g., reproductive toxins, select carcinogens, and chemicals with a high degree of acute toxicity.

Reproductive toxins are chemicals that affect the reproductive capabilities of either male or female including chromosomal damage (mutagenesis) and effects on fetuses (teratogenesis).

Select carcinogens are chemicals that meet any one of the following criteria:

(1) Any chemical regulated by OSHA as a carcinogen;
(2) Any chemical listed in the Annual Report on Carcinogens under the category of "known carcinogens" published by the National Toxicology Program (NTP);
(3) Any chemical listed under Group 1 (carcinogenic to humans) by the International Agency for Research on Cancer (IARC); or
(4) Any chemical listed in either Group 2A or 2B by IARC or under the category "reasonably anticipated to be carcinogens" by NTP, that may cause statistically significant tumor incidence in experimental animals in accordance with any of the following exposure conditions: (a) after inhalation exposure of 6 to 7 hr per day, 5 days per week, for a significant portion of a lifetime to dosages of less than 10 mg/M^3; (b) after repeated skin absorption of less than 300 mg/kg of body weight per week; and (c) after oral dosage of less than 50 mg/kg of body weight per day.

A chemical with a high degree of acute toxicity is any substance for which the LD_{50} data cause the substance to be classified as a "highly toxic chemical" as defined by the American National Standards Institute (ANSI Z129.1).

If reproductive toxins, select carcinogens, or chemicals of a high degree of acute toxicity are used in the laboratory, the employer must identify and post one or more areas as *designated areas*. A designated area may be the entire laboratory, an area of a laboratory, or a device such as a laboratory fume hood or a glove box. Once a designated area is posted, the boundaries of that area must be clearly marked. Only trained workers are to be permitted to work in that area.

5 GENERAL SAFETY PRACTICES

The most important safety rule in any laboratory is that every person involved in laboratory operations, from administrator to individual worker, must be safety minded. It is impossible to design a set of safety rules that will cover all possible hazards and occurrences. Safety awareness can become part of everyone's habits only if safety issues are discussed repeatedly and only if senior and responsible staff evince a sincere and continuing interest and are perceived by all their associates as doing so. The laboratory worker, however, must accept responsibility for carrying out his or her own work in accordance with good safety practices.

Laboratory supervisors have the overall safety responsibility and must provide for scheduled formal safety inspections as well as continual informal inspections. They have the responsibility to ensure that:

(1) Laboratory workers know safety rules and follow them;
(2) An appropriate safety orientation has been given to individuals when they are first assigned to a laboratory space;
(3) Information on special or unusual hazards in nonroutine work has been disseminated to the operator;
(4) Adequate emergency equipment in proper working order is available; and
(5) Training in the use of emergency equipment has been provided.

Too much familiarity with a particular laboratory operation may result in overlooking or underestimating its hazards. Such an attitude can lead to a false sense of security, which frequently results in carelessness. Every laboratory worker has a basic responsibility to himself or herself and colleagues to plan and execute laboratory operations in a safe manner. Advance planning is one of the best ways to avoid serious incidents.

25

Some general guidelines for safety practices in the laboratory are given below that experience has shown to be useful for avoiding accidents or reducing injuries in the laboratory.

GENERAL SAFETY RULES

In addition to any specific safety rules for some operations, every laboratory worker is to observe the following general safety rules:

(1)   Know and follow the safety rules and procedures that apply to the operation.

(2)   Use the proper type of personal protective apparel and equipment for the operation.

(3)   Be alert to unsafe conditions and reactions. Call attention to hazards so that corrections can be made immediately.

(4)   Use laboratory equipment only for its designed purpose.

(5)   Know the location of and how to use the emergency equipment in your area, as well as how to obtain additional help in an emergency. Read and become familiar with emergency response procedures.

(6)   Remain out of the area of a chemical spill or accident unless it is your responsibility to help meet the emergency. Bystanders interfere with emergency response and rescue and endanger themselves.

(7)   Avoid consuming food or beverages or smoking in any area where chemicals are being used or stored.

(8)   Avoid distracting or startling any co-worker in the laboratory. Practical jokes or horseplay cannot be tolerated at any time in the laboratory.

(9)   Label all chemicals clearly and correctly.

(10) Post warning signs when unusual hazards, such as ultraviolet (UV) irradiation, pressurized reaction, or other special safety problems may exist.

(11) Access to exits, emergency equipment, controls, and such must never be blocked. Stairways and hallways must not be used as storage areas even temporarily.

(12) Laboratory equipment must be inspected regularly and serviced according to the manufacturer's suggested schedule. Maintenance plans must include a procedure to ensure that an out-of-service piece of equipment cannot be restarted.

(13) Whenever a potential explosion or implosion might occur, a topple-resistant safety shield for personnel protection must be used.

(14) Operations known to be hazardous must not be undertaken by a worker working alone in a laboratory.

(15) Emergency telephone number to be called in the event of a chemical accident or emergency must be posted prominently in each laboratory. In addition, the telephone numbers of the laboratory workers and their supervisors must also be posted in the event of an accident or emergency.

PERSONAL HYGIENE AND SAFETY RULES

Every laboratory worker must observe the following personal hygiene and safety rules:

(1) All laboratory workers must wear appropriate personal protective apparel and equipment as well as footwear while in the laboratory.
(2) Avoid wearing loose clothing and confine long hair when in the laboratory.
(3) Avoid mouth suction to pipet chemicals or to start a siphon. A pipet bulb or an aspirator should be used to provide vacuum.
(4) Before leaving the laboratory area, workers must wash well with soap and water. Under no circumstances should any laboratory worker wash with organic solvents.
(5) Avoid applying cosmetics in any area where chemicals are being used or stored.
(6) Laboratory refrigerators, cold rooms, ice chests, or similar laboratory equipments must not be used for food storage. Separate equipment should be dedicated to that use and prominently labeled.
(7) Glassware or laboratory utensils that have been used for laboratory operations must never be used to prepare or consume food or beverages.
(8) Personal laboratory space or work area must be kept clean and free from obstructions. Cleanup must follow the completion of any operation or at the end of each work day.
(9) Any spilled chemicals must be cleaned up immediately and disposed of properly in accordance with established laboratory waste management procedures.

LABORATORY GLASSWARE SAFETY RULES

A leading cause of laboratory injuries is due to accidents involving glassware. Every laboratory worker must observe the following safety rules involving glassware:

(1) Avoid damaging any laboratory glassware by following careful handling and storage procedures. Discard any damaged glassware items into proper containers.

(2) Hands are to be properly protected when inserting glass tubing into rubber stoppers or corks or when placing rubber tubing on glass hose connections. Glass tubing must be fire-polished or rounded and lubricated, and hands must never be held close together to limit movement of glass should it break or fracture.

(3) Handle vacuum-jacketed glass apparatus with extreme care to prevent implosion. Dewar flasks must be properly taped. If vacuum work is performed, only glassware designed for such purpose may be used.

(4) Workers must be properly instructed in the use of glass equipment designed for specialized operations.

(5) Hands must be properly protected when picking up any broken glass. Small pieces must be swept up with a brush into a dustpan.

HOUSEKEEPING SAFETY RULES

It has been observed time and time again that there is a definite relationship between safety record and orderliness in the laboratory. When housekeeping is poor, safety record also deteriorates. The laboratory work area must be kept clean and orderly, and chemicals and equipment must be properly labeled and stored. Every laboratory worker must observe the following housekeeping rules:

(1) All work areas, especially laboratory benches, must be kept clear of clutter.

(2) All working surfaces and floors must be cleaned regularly.

(3) Access to all emergency equipment, showers, eyewashes, and exits must never be blocked by anything, not even by a temporarily parked chemical cart.

(4) No chemicals or equipment are to be stored in aisles or stairwells, in fume hoods, on laboratory benches or desks, or floors or in hallways, or to be left overnight on shelves over the workbenches.

(5) All chemical containers must be properly labeled with the identity of the contents and the hazards present to users.

(6) At the end of each work day, all chemicals must be placed in their assigned storage areas, and the contents of all unlabeled containers must be considered wastes that are to be properly placed in labeled containers.

(7) All chemical spills and accidental releases must be promptly cleaned up, in accordance with the chemical emergency response plan, using appropriate protective apparel and equipment, and the spilled chemical and cleanup materials properly disposed of.

MISCELLANEOUS SAFETY RULES

In addition to the general safety practices already discussed, every laboratory worker must observe the following safety practices as well:

Unattended Operations

Sometimes, it is necessary to carry out a laboratory operation continuously or overnight. It is essential for the operator to plan for interruptions in utility services such as cooling water, electricity, and inert gas. A contingency plan must be made to avoid hazards in case of utility failure. The operator needs to arrange for routine inspection of the operation. For operations running overnight, the laboratory lights must be left on and an appropriate sign and information on contact person must be placed prominently on the laboratory door.

Working Alone

Generally, it is prudent to avoid working in a laboratory alone. The laboratory supervisor has the responsibility to determine whether the operation requires special safety precautions such as having two workers in the same laboratory when a particular operation is being carried out. Even if the laboratory supervisor decides that an operator can work alone, arrangements must still be made to have another operator working in a nearby laboratory within earshot distance to cross-check periodically.

Cryogenic Hazards

The primary hazard of cryogenic materials is their extreme coldness. They can cause severe burns if allowed to contact the skin. Appropriate dry gloves, face shield and protective apparel must be worn when handling cryogenic materials. Neither liquid nitrogen nor liquid air should be used to cool a flammable mixture in the presence of air because oxygen can condense from the air and may result in an explosion. In the preparation of a cooling bath, dry ice, which is solid carbon dioxide, must be added slowly to the liquid portion to avoid foaming over.

Pressurized Systems

Pressurized systems must have an appropriate relief device. If the reaction cannot be opened directly to the air, an inert gas purge and bubbler system must be used to avoid pressure buildup. Reactions must never be carried out in, nor heat applied to, an apparatus that is a closed system unless it is designed and tested to withstand pressure.

Mechanical Equipment

All mechanical equipment must be adequately furnished with guards that prevent access to electrical connections or moving parts such as belts and pulleys of a vacuum pump. Each laboratory worker must inspect equipment before using it to ensure that the guards are in place and functioning. A defective or ineffective guard can be worse than none at all, because it can give a false sense of security. In addition to mechanical and electrical guarding, emergency shutoff devices may be needed.

6 PERSONAL PROTECTIVE EQUIPMENT

The objective of a personnel protection program is to protect laboratory workers from health hazards of chemicals as well as physical hazards and to prevent injury to the user from the incorrect use and/or malfunction of safety equipment. All laboratory workers must be protected from potential chemical and physical hazards by personal protective apparel and equipment. Every worker must be fully informed of the shielding capabilities as well as the limitations of their personal protective apparel and equipment and be trained in their proper use. A comprehensive personnel protection program shall contain the following components:

(1) *Hazard Assessment.* The selection and purchase of personal protective apparel and equipment is to be based upon an assessment of anticipated hazards in the work area.

(2) *Medical Evaluation.* Medical evaluation of personnel for fitness of duty and use of personnel protective equipment must be one of the criteria for selection and continuing certification of its use.

(3) *Equipment Selection and Use.* The relationship between the anticipated hazard, the environment being encountered, and the personal protective equipment ensemble must be recognized. In some instances, the weakest link in the system may be the physical and mental condition of the user.

(4) *Employee Training.* An effective personnel protection program cannot exist without a mandatory and comprehensive employee training program.

(5) *Equipment Inspection, Maintenance, and Storage.* Regular inspection, maintenance, and proper storage are key elements in any personnel protection program.

Table 6-1   OSHA Standards on Personnel Protection

OSHA Standard	Type of Protection
29 CFR 1910.95	Hearing Protection
29 CFR 1910.132	General Protection
29 CFR 1910.1000	
29 CFR 1910.1001-1045	
29 CFR 1910.133	Eye and Face Protection
29 CFR 1910.134	Respiratory Protection
29 CFR 1910.135	Head Protection
29 CFR 1910.136	Foot Protection

OSHA comprehensively regulates employee health and safety. A written personnel protection program including the aforementioned components is required by OSHA. The written program must also include a policy statement with the guidelines and procedures listed above. Copies of the written program must be made available to all employees or their designated representative. Technical data concerning equipment, maintenance manuals, relevant regulations, and other essential information must also be made available if requested. Table 6-1 summarizes current OSHA regulations on use of personal protective equipment. The personnel protection program must be reviewed annually and the results are to be made available to employees and presented to top management so that improvements may be made and implemented.

A wide variety of specialized protective clothing and safety equipment is commercially available for use in the laboratory. The proper use of these items will minimize or eliminate exposure to the hazards associated with many laboratory operations. The primary goal of laboratory safety procedures is the prevention of accidents and emergencies. However, accidents and emergencies may nonetheless occur and, at such times, proper safety equipment and correct emergency response procedures can help minimize injuries or damage.

EYE AND FACE PROTECTION

Eye protection must be required for all laboratory workers and any visitors present in locations where chemicals are used or stored. No one should be allowed to enter any laboratory without appropriate eye protection. Ordinary prescription glasses do not provide adequate protection from injury to the eyes. The minimum acceptable eye protection requires the use of hardened-glass or plastic safety spectacles. Safety spectacles that meet the criteria described below provide minimum eye protection for regular use. Laboratory management must make appropriate eye-protection devices available to visitors or others who only occasionally enter eye-protection area. It is important that each laboratory operation be analyzed to ensure that adequate eye and face protection is used. It is the responsibility of the laboratory supervisor to determine the level of eye and face protection required and to enforce safety rules.

Contact Lenses

Contact lenses are not to be worn in a laboratory. Gases and vapors can be concentrated under such lenses and cause permanent eye damage. Furthermore, in the event of a chemical splash into an eye, it is often nearly impossible to remove the contact lens to irrigate the eyes because of involuntary spasm of the eyelid. Persons attempting to irrigate the eyes of an unconscious victim may not be aware of the presence of contact lenses, thus reducing the effectiveness of such treatment. Soft lenses can absorb solvent vapors even through face shields and, as a result, adhere to the eye. In the event that contact lenses must be worn for medical or therapeutic reasons, the individuals must inform the laboratory supervisor so that proper safety precautions can be devised.

Safety Spectacles

Safety spectacles used in the laboratory must comply with the Standard for Occupational and Educational Eye and Face Protection established by the American National Standards Institute (ANSI Z87.l). This standard specifics a minimum lens thickness of 3 mm; impact resistance requirements; passage of a flammability test; and lens-retaining frames.

Side Shields

Side shields that attach to regular safety spectacles offer some protection from objects that approach from the side but do not provide adequate protection from chemical splashes. Other eye and face protection must be worn when a significant chemical splash hazard exists.

Goggles

Goggles are intended for wear when there is danger of flying particles or splashing of chemicals. Impact-protection goggles have perforation on the sides to provide ventilation and reduce fogging of the lenses and thus do not offer full protection against chemical splashes. Splash goggles, also known as "acid goggles", do not have perforation on the sides and their use is required when protection from harmful chemical splash is warranted.

Face Shields

Goggles provide protection to the eyes but offer little protection to the face and neck. Full-face shields that protect the face and throat must always be worn when maximum protection from flying particles and harmful chemical splashes is needed. For full protection, safety glasses must be worn with face shields. A full-face shield must be worn when an operation involves a pressurized system as it may either explode or implode.

Specialized Eye Protection

There are specific goggles and masks for protection against radiation and laser hazards, as well as intense light sources. The laboratory supervisor must determine whether the task being performed requires specialized eye protection and enforce its use.

BODY PROTECTION

The clothing worn by laboratory workers can be important to their safety. Laboratory workers must not wear loose, torn, or skimpy clothing. Loose or torn clothing items are fire hazards. They may dip into chemicals or become ensnared in apparatus or moving parts of laboratory equipment. Skimpy clothing offers little protection to the skin in the event of chemical splash.

Appropriate personal protective apparel is advisable for most laboratory work and may be required for some. If the possibility of chemical contamination exists, personal clothing that will be worn home should be covered by protective apparel. Such protective apparel can include laboratory coats and aprons, or jumpsuits. In chemical emergency response situations, impervious chemical suits fully enclosing the body of the responder may be necessary.

Personal protective apparel can be either washable or disposable. It should permit the wearer easy execution of manual tasks while being worn. In addition, it should also satisfy other performance requirements such as strength, chemical resistance, flexibility, and ease of cleaning. Garments are commercially available that can help protect the laboratory worker against minor chemical splashes, heat, cold, moisture, and radiation. The choice of garment — laboratory coat vs rubber or plastic apron vs jumpsuits — depends on the degree of protection required and is the responsibility of the laboratory supervisor.

Laboratory Coats

Laboratory coats are intended to prevent contact with dirt and the minor chemical splashes or spills encountered in laboratory work. The cloth laboratory coat is primarily a protection for clothing. Sometimes, it may itself present a hazard to the wearer. Cotton and synthetic materials such as Tyvek are satisfactory. Rayon and polyester are not. Laboratory coats do not significantly resist penetration by organic liquids and, if significantly contaminated by them, should be removed immediately.

Plastic or Rubber Aprons

Plastic or rubber aprons provide better protection from corrosive or irritating liquids but can complicate injuries in the event of fire. It should be noted that a plastic apron can accumulate a considerable charge of static electricity and must be avoided in areas where flammable solvents or other materials could be ignited by a static discharge.

Jumpsuits

Jumpsuits are strongly recommended for high-risk situations. They are made of disposable material such as Tyvek. An example of a high-risk situation is the handling of appreciable quantities of carcinogenic chemicals, for which long sleeves and the use of gloves and head and shoe covers are also recommended. Many disposable garments, however, offer only limited protection from vapor penetration and considerable judgment is needed when using them.

HAND PROTECTION

Skin contact is a potential problem and source of exposure to corrosive or toxic chemicals. Gloves must be worn whenever it is necessary to handle corrosive or toxic chemicals, broken or damaged glassware, rough or sharp-edged objects, hot or cold materials, or whenever protection is needed against unintentional exposure to chemicals. Gloves are to be selected on the basis of the material being handled, the particular hazard involved, and their suitability for the operation being conducted. Before each use, they must be inspected for punctures, tears, or discoloration. Before removal, they must be washed appropriately in accordance with the manufacturer's recommendation. Glove materials are eventually permeated by chemicals. However, they can be used safely for limited time periods if specific use and glove characteristics, such as thickness and permeation rate and time, are known. Some of this information can be obtained from the manufacturer, or the gloves used can be tested for breakthrough rates and times by the user. Gloves must be replaced periodically, depending on frequency of use and permeability to the chemicals handled.

Rubber Gloves

There are various compositions and thicknesses of rubber gloves. Common rubber glove materials include butyl or natural rubber, neoprene, nitrile, and polyvinyl chloride. These materials differ in their resistance to various substances and this information is summarized in Table 6-2. Rubber gloves are to be inspected before each use. Periodically, an inflation test, in which the glove is first inflated with air and then immersed in water and examined for the presence of air bubbles, must be conducted.

Leather Gloves

Leather gloves may be used for handling broken glassware, for inserting glass tubes into rubber stoppers, and for similar operations where protection from chemicals is not needed.

Insulated Gloves

Insulated gloves must be worn when working at temperature extremes. In the past, many insulated gloves were made either entirely or partly of asbestos. It is best to dispose of such gloves if they are still around because asbestos is regulated as a carcinogen under OSHA.

RESPIRATORY PROTECTION

The OSHA General Industry Standard for respiratory protection (29 CFR 1910.134) requires that a respiratory protection program be established by the employer, and that respirators are provided and are effective when such equipment is necessary to protect the health of the worker. However, OSHA health standards place primary emphasis on engineering, administrative, or work practice controls in light of the inherent deficiencies of respirators and respiratory programs.

The employer must have written standard operating procedures governing the selection and use of respirators. The procedures must include a discussion or explanation of all the items specified in the respiratory protection standard. Determining the adequacy of the written procedure is a professional judgment made by an OSHA compliance officer at the time of visit.

**Table 6-2   Chemical Resistance of Common Glove Materials
(E: Excellent, G: Good, F: Fair, P: Poor)**

Chemical	Natural Rubber	Neoprene	Nitrile	Vinyl
Acetaldehyde	G	G	E	G
Acetic Acid	E	E	E	E
Acetone	G	G	G	F
Acrylonitrile	P	G	—	F
Ammonium Hydroxide	G	E	E	E
Aniline	F	G	E	G
Benzaldehyde	F	F	E	G
Benzene[a]	P	F	G	F
Benzyl Chloride[a]	F	P	G	P
Bromine	G	G	—	G
Butane	P	E	—	P
Butylaldehyde	P	G	—	G
Calcium Hypochlorite	P	G	G	G
Carbon Disulfide	P	P	G	F
Carbon Tetrachloride[a]	P	F	G	F
Chlorine	G	G	—	G
Cloroacetone	F	E	—	P
Chloroform[a]	P	F	G	P
Chromic Acid	P	F	F	E
Cyclohexane	F	E	—	P
Dibenzyl Ether	F	G	—	p
Dibutyl Phthalate	F	G	—	P
Diethanolamine	F	E	—	E
Diethyl Ether	F	G	E	P
Ethyl Acetate	F	G	G	F
Ethylene Dichloride[a]	P	F	G	P
Ethylene Glycol	G	G	E	E
Ethylene Trichloride[a]	P	P	—	P
Fluorine	G	G	—	G
Formaldehyde	G	E	E	E
Formic Acid	G	E	E	E
Glycerol	G	G	E	E
Hexane	P	E	—	P

Table 6-2   (Continued)
Chemical Resistance of Common Glove Materials
(E: Excellent, G: Good, F: Fair, P: Poor)

Chemical	Natural Rubber	Neoprene	Nitrile	Vinyl
Hydrobromic Acid	G	E	—	E
Hydrochloric Acid	G	G	G	E
Hydrofluoric Acid	G	G	G	E
Hydrogen Peroxide	G	G	G	E
Iodine	G	G	—	P
Methyl Cellosolve	F	E	—	P
Methyl Chloride[a]	P	E	—	P
Methyl Ethyl Detone	F	G	G	P
Methylamine	G	G	E	E
Methylene Chloride[a]	F	F	G	F
Monoethanolamine	F	E	—	E
Morpholine	F	E	—	E
Naphthalene[a]	G	G	E	G
Nitric Acid	P	P	P	G
Perchloric Acid	F	G	F	E
Phenol	G	E	—	E
Phosphoric Acid	G	E	—	E
Potassium Hydroxide	G	G	G	E
Propylene Dichloride[a]	P	F	—	P
Sodium Hydroxide	G	G	G	E
Sodium Hypochlorite	G	P	F	G
Sulfuric Acid	G	G	F	G
Toluene[a]	P	F	G	F
Trichloroethylene[a]	P	F	G	F
Tricresyl Phosphate	P	F	—	F
Triethanolamine	F	E	E	E
Trinitrotoluene	P	E	—	P

[a] Aromatic and halogenated hydrocarbons will attack all types of natural and synthetic glove materials.

Under OSHA regulations, only equipment listed and approved by the Mine Safety and Health Administration (MSHA) and the National Institute for Occupational Safety and Health (NIOSH) may be used for respiratory protection.

Approved numbers are clearly written on all approved respiratory equipment. However, not all respiratory equipment currently in the market is approved. NIOSH periodically publishes a list, entitled *NIOSH Certified Equipment List,* of all approved respirators and respiratory equipment.

The level of protection that can be provided by a respirator is indicated by the *protection factor* which is a number. This number, determined experimentally by measuring facepiece seal and exhalation valve leakage, indicates the relative difference in concentrations of chemical outside and inside the facepiece that can be maintained by the respirator. For example, if the protection factor for a certain type of respirator is 50, this means that workers wearing this type of respirator should be protected in atmospheres containing chemicals at concentrations that are up to 50 times higher than the approximate limits. Protection factors for various types of respirators are available from the manufacturers. However, due to individual differences and personal habits, the protection factor for a specific type of respirator must be determined experimentally for each worker.

The OSHA respiratory protection standard requires that the respirator user be instructed and trained in the proper use and inspection of respirators as well as their limitations. This includes requirements for demonstrations on how the respirator should be worn, how to adjust it, and how to determine if it fits properly. It is the responsibility of the employer to see to it that the respirators are regularly and properly cleaned, disinfected, and stored in a clean and sanitary location. Workers are not to use respirators unless it has been determined by a licensed physician that they are medically fit and physically able to perform the work while wearing the respirator.

Respirator training must provide the opportunity for the worker to handle the respirator, have it fitted properly, test its facepiece-to-face seal, wear the respirator in normal air, and finally, to wear it in a test atmosphere. This *quantitative* respirator fit test involves exposing the respirator wearer to a test atmosphere containing an easily detectable, relatively nontoxic aerosol, vapor, or gas as the test agent and measuring the penetration of the agent into the respirator via odor detection of the wearer. *Qualitative* respirator fit test involves a wearer's responding to a chemical outside the respirator facepiece. Qualitative fit tests are fast, easily performed, and use inexpensive equipment. Some commonly used qualitative fit tests include irritant smoke test (stannic chloride or titanium chloride), odorous vapor test (isoamyl acetate), and taste test (sodium saccharin).

Respirators shall not be worn when facial conditions prevent a good face seal. Such conditions may include beard growth, sideburns, caps or hats that project under the facepiece, or temple bars on eyeglasses. Wearing of contact lenses is prohibited in contaminated atmospheres when using a respirator.

Respirators for emergency use must be inspected after each use and at least monthly to assure that they are in satisfactory working condition. Self-contained breathing apparatus (SCBA) must be inspected monthly. The inspections must ensure that cylinders are fully charged and that regulator and wearing devices are properly functioning. Records must be kept of inspection dates and findings for respirators maintained for emergency use.

Several types of nonemergency respirators are available for protection in atmospheres that are not immediately dangerous to life or health but could be detrimental after prolonged or repeated exposure. Other types of respirators are available for emergency or rescue work in atmospheres from which the wearer cannot escape without respiratory protection. The choice of the appropriate respirator to use in a given situation will depend on the type of contaminant and its estimated or measured concentration, known exposure limits, and warning and hazardous properties. Once this information is available, an appropriate type of respirator can be selected by following the flow chart in Figure 6-1.

A respiratory protection program can be successful only if the management can gain employee acceptance. Many factors affect the employee's acceptance of respirators, including comfort, ability to breathe without objectionable effort, adequate visibility under all conditions, provisions for wearing prescription glasses if necessary, ability to communicate, ability to perform all tasks without undue interference, and confidence in the facepiece fit. Failure to consider these factors is likely to reduce cooperation of the users in promoting a satisfactory program. How well these problems are resolved can be determined by management observation of respirator users during normal activities and management solicitation of employee comments.

Air-Purifying Respirators

Air-purifying respirators (APR) consist of a facepiece and an air-purifying device that is either a removable component of the facepiece or an air-purifying apparatus worn on a body harness and attached to the facepiece by a corrugated breathing hose. These respirators range from full facepiece, dual cartridge to those with no eye protection. Air-purifying respirators can be used only for protection against either a particular gas or vapor, or a class of vapors or gases as specified by the respirator manufacturer. They cannot be used at concentrations of contaminants above that specified on the cartridge. In addition, these respirators cannot be used if: the oxygen content of the air is less than 19.5%; in atmospheres with airborne concentrations immediately dangerous to life and health (IDLH); or for chemical emergency response and rescue work. They are also not to be used in areas where airborne concentrations are unknown or exceed the "designated use concentrations," which are based upon testing at a given temperature over a narrow range of flow rates and relative humidities. Because it is possible for significant breakthrough to occur at a fraction of the cartridge capacity, knowledge of the potential workplace exposure and length of time the respirator will be worn is important. It may be desirable to replace the cartridge after each use to ensure the maximum available exposure time for each new use. Difficulty in breathing may indicate plugged or exhausted filters or cartridges.

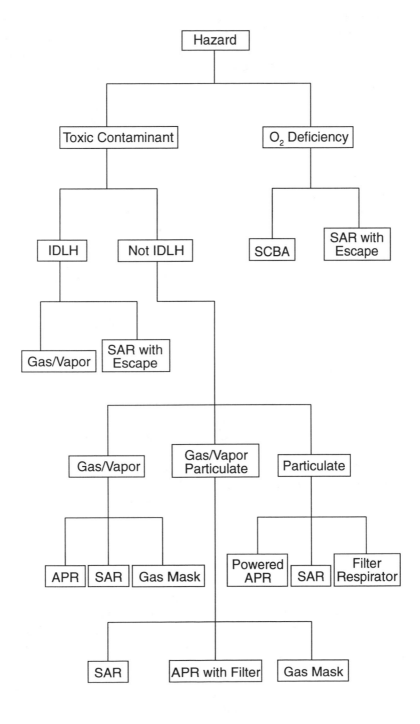

Figure 6-1   Flow chart for respirator selection.

Detection of odors may indicate concentrations of contaminants higher than the absorbing or adsorbing capacity of the cartridge. In either case, the air-purifying respirator should be replaced immediately. Because most air-purifying respirators are of negative pressure design, OSHA requires all users of these respirators to be individually fit-tested for an adequate facepiece seal.

Supplied-Air Respirators

Supplied-air respirators (SAR), also referred to as air-line respirators, have an air supply from a source located some distance away and connected to the user by an air-line hose. Supplied-air respirator and other atmosphere-supplying respirators, such as SCBA, provide the highest available level of protection against airborne contaminants and in an oxygen-deficient atmosphere. Various types of facepieces are available and they include: the *half masks,* which cover the face from below the chin to over the nose but do not provide eye protection; the *full-facepiece masks,* which cover the face from the hairline to below the chin and provide eye protection.

In the continuous-flow mode, these respirators supply fresh air, never oxygen, to the facepiece of the respirator at a pressure high enough to cause a slight buildup relative to atmospheric pressure. As a result, the supplied air flows outward from the mask and contaminated air from the work environment cannot readily enter the facepiece. The air supply of this type of respirator must be kept free of contaminants, e.g., by use of oil filters and carbon monoxide absorbers, and some consideration should be given to its quality and relative humidity.

Supplied-air respirators are not approved for use in atmospheres immediately dangerous to life and health or in oxygen-deficient atmospheres unless they are equipped with an emergency escape unit such as an escape-only self-contained breathing apparatus which provides immediate emergency respiratory protection in case of air-line failure. All supplied-air respirators must have couplings that are incompatible with the outlets of other gas systems used on site to prevent accidental connection to an inappropriate compressed gas source. Supplied-air respirators suitable for use with compressed air are classified as "Type C" respirators as defined by the Mine Safety and Health Administration and the National Institute for Occupational Safety and Health (MSHA/NIOSH).

Using an air-line hose requires the user to drag long lengths of hose connected to the air supply. Thus, the range of use is limited to the maximum length of the hose and the worker must retrace steps to leave the work area. As the hose length is increased, the minimum approved air flow may not be delivered at the facepiece. The maximum hose length from the source to the respirator is limited by NIOSH regulations to 300 ft. It should be noted that the air-line hose is vulnerable to physical damage, chemical contamination, degradation, and the supplied-air respirator user also requires monitoring or supervision of the air supply line.

Self-Contained Breathing Apparatus

Self-contained breathing apparatus (SCBA) is the only type of respiratory equipment that offers protection against most types of airborne contaminants, including concentrations that are immediately dangerous to life and health (IDLH), or oxygen deficiency. This equipment consists of a full-face mask connected to a cylinder of compressed air. However, the air supply is limited to the capacity of the cylinder and its rate of consumption by the user. Therefore, these respirators cannot be used for extended periods without recharging or replacing the cylinder.

Mine Safety and Health Administration and National Institute for Occupational Safety and Health (MSHA/NIOSH) requires that all compressed breathing gas cylinders must meet minimum requirements of the Department of Transportation (DOT) for interstate shipment. All compressed air, compressed oxygen, liquid air, and liquid oxygen used for respiration shall be of high purity and must meet all requirements of the OSHA respiratory protection standard (29 CFR 1910.134). In addition, breathing air must meet or exceed the requirements of Grade D breathing air as specified in the Compressed Gas Association Pamphlet G-7.1 and by the American National Standards Institute (ANSI Z86.1-1973).

All institutions or organizations with laboratories in which chemicals are used must have protective equipment of this type available for emergencies and provide training in its use. Under current regulations of the Mine Safety and Health Administration, self-contained breathing apparatus is approved either for escape only or for both entry into and escape from a hazardous atmosphere.

Escape-Only SCBA

Escape-only SCBA supplies clean air to the user from either an air cylinder or from an oxygen-generating chemical. Frequently, they are continuous-flow devices with hoods that can be donned to provide immediate emergency protection. The chief advantage of the escape-only SCBA is that it is lightweight (weighing 10 lb or less), low bulk, and easy to carry. However, they only provide 5 to 15 min of respiratory protection, depending on the model and user's rate of consumption, and cannot be used for entry.

Entry-and-Escape SCBA

The entry-and-escape SCBA supplies oxygen from a source or air from a cylinder. It allows the user untethered access to nearly all areas of the work site but limits the user's mobility due to both the bulk and weight of the unit. Two types of entry-and-escape SCBA are available: open-circuit SCBA and closed-circuit SCBA (or rebreather).

The *open-circuit SCBA* supplies fresh air to the facepiece regulator from a compressed air cylinder worn on a backpack. Exhaled air is discharged to atmosphere directly outside the face mask. The open-circuit SCBA units weigh about 35 lb and, depending on the model, provide an average of either 30 or 45 min of breathing air supply. Obviously, workers under stress will consume the unit's rated air supply more rapidly. In all units, a warning alarm sounds when only 20 to 25% of the air supply remains.

The *closed-circuit SCBA* or rebreather recycles exhaled air and exhaled carbon dioxide is removed by the use of a disposable alkaline scrubber. Oxygen is replenished into the breathing air from a liquid oxygen tank mounted on the unit's backpack. Its advantages, over an open-circuit SCBA, are longer operating time of up to 4 hr and a lighter weight of 20 to 30 lb. Again, warning alarm sounds when only 20 to 25% of the oxygen supply remains and oxygen supply is depleted before the carbon dioxide sorbent scrubber supply. A major disadvantage of a closed-circuit SCBA is that the unit's breathing bag must never be worn outside an encapsulating suit because the bag may be permeated by chemicals, contaminating the apparatus and the breathing air. Decontamination of the breathing bag may also be impossible. At very cold temperatures, the alkaline scrubber efficiency may be reduced and carbon dioxide breakthrough may occur. Both the heat exhaled by open-circuit SCBAs and that generated in the carbon dioxide scrubber are retained in the unit, possibly adding to the individual's heat stress.

STANDARD SAFETY AND EMERGENCY EQUIPMENT

Additional safety and emergency equipment must be available in all laboratories. Laboratory workers must realize that such equipment is intended to help protect them from potential injury and need to be trained in its proper use. Every laboratory worker must learn the locations of and how to use the available safety and emergency equipment in their work area so that he or she can find them with eyes closed, if necessary. All laboratories in which chemicals are used are to have available eyewash fountains, safety showers, fire extinguishers, and other safety and emergency equipment depending on the operations performed. It is the responsibility of the chemical hygiene officer to recommend and provide supplementary safety and emergency equipment as needed. It is further the responsibility of the chemical hygiene officer to ensure that all safety and emergency equipment are routinely inspected, maintained, and in good working condition consistently, for each and every use.

Eyewash Fountains

Eyewash fountains must be available in all laboratories where chemicals are being used or stored. An eyewash fountain must provide a soft stream or spray of aerated water for a period of at least 15 min. Eyewash fountains must be located close to the safety showers so that, if necessary, the eyes can be washed within 10 seconds while the body is showered.

Safety Showers

Safety showers must be installed in laboratory areas where chemicals are handled for immediate first aid treatment of chemical splashes and for extinguishing clothing fires. Safety showers are to be tested routinely by laboratory personnel to ensure that all valves are operable and to remove any debris that may have accumulated in the system. Safety showers should be capable of drenching the subject immediately and must be large enough to accommodate more than one person if necessary. A safety shower must have a quick-opening valve that requires manual closing. A downward-pull delta bar is satisfactory, if long enough, but chain pulls are not advisable because of the difficulty of grasping them in an emergency (ANSI Z-358.1).

Fire Extinguishers

All chemical laboratories should be provided with carbon dioxide, dry chemical, or other fire extinguishers. Other types of fire extinguishers must also be made available if required by the operation being performed in the laboratory. The four types of fire extinguishers most commonly used are classified by the type of fire for which they are intended.

Water extinguishers are effective against burning paper and ordinary combustibles (Class A fires). They should not be used for extinguishing electrical, burning liquid, or metal fires.

Carbon dioxide and halon extinguishers are effective against burning liquids, such as hydrocarbons or solvents, and electrical fires (Classes B and C fires). They are recommended for fires involving delicate instruments and electronic systems because they do not damage such equipment. The are not effective against paper and ordinary combustibles or metal fires. It should be noted that carbon dioxide extinguishers should not be used against lithium aluminum hydride fires.

Dry powder extinguishers contain sodium bicarbonate. They are effective against burning liquids and electrical fires (Classes B and C fires) but are not effective against paper and ordinary combustible or metal fires. They are not recommended for fires involving delicate instruments, electronic systems, or optical systems because of the cleanup problem after their use.

Met-L-X extinguishers and others that have special granular formulations are effective against burning metal (Class D fires). Included in this category are fires involving magnesium, potassium, lithium, and sodium; alloys of reactive metals; metal hydrides; metal alkyls; and other organometallic compounds. These extinguishers are not effective against paper and ordinary combustibles, burning liquids, or electrical fires.

Every fire extinguisher must carry a label indicating what class or classes of fires is it effective against. Each laboratory worker must know the location and limitation of the fire extinguishers in the work area and be trained in their proper use. OSHA requires annual hands-on training if employees are expected to use any type of the fire extinguishers. After use, a fire extinguisher must either be recharged or be replaced by the chemical hygiene officer.

7   IDENTIFICATION OF HAZARDOUS CHEMICALS

In the U.S., the responsibility for identifying and regulating hazardous chemicals lies with several government agencies. The Environmental Protection Agency (EPA) regulates the use and labeling of agricultural chemicals and pesticides. The Food and Drug Administration (FDA) regulates the manufacture and use of pharmaceutical products. The Department of Transportation (DOT) regulates the interstate transportation and marking of hazardous materials, including those regulated by the EPA and the FDA. The type of vehicle, the origin, and the destination of the materials determine which jurisdiction (local, state, federal, or international) and dictates the conditions of shipment. The United Nations (UN) has established guidelines for identifying hazardous materials that one transports from one country to another. A country using the UN system requires that hazardous chemicals coming into the country be properly placarded and labeled.

THE UNITED NATIONS NUMBERING SYSTEM

The UN has developed a two-part system that the U.S. and Canada use in conjunction with illustrated placards. The UN system provides uniformity in recognizing hazardous materials in international transport. The first part of the UN numbering system divides hazardous materials into nine hazard classes numbered 1 through 9 (Table 7-1). This general hazard class number appears at the bottom of the DOT placard.

The second part of the UN system identifies the transported materials by assigning a four-digit number to specific materials and is required on tank cars, cargo tanks, and portable tanks only. This number may appear across the face of a standard DOT placard or on its own orange rectangular panel. In the U.S., the

Table 7-1   United Nations Hazard Classes

Hazard Class	Materials	Includes
1	Explosives	"A", "B", "C", blasting agents
2	Gases	Flammable, nonflammable, poison gases
3	Flammable liquids	Flammable, combustible liquids
4	Flammable solids	Flammable, combustible, water-reactive solids
5	Oxidizers	Oxidizers, peroxides
6	Poisons	Gases, liquids, solids, etiologic agents
7	Radioactive materials	I, II, III
8	Corrosives	Acids, alkalis
9	Other hazardous materials not otherwise classified	Consumer goods, hazardous waste

usage of the separate orange UN number panel is not permitted without also using the DOT illustrated hazard placard. Examples of specific materials, their UN hazard class, and their four-digit identification numbers are shown in Table 7-2. A good reference source for the UN four-digit identification numbers is the *Emergency Response Guide* of the U.S. Department of Transportation.

THE U.S. DEPARTMENT OF TRANSPORTATION SYSTEM

The U.S. Department of Transportation regulates the transportation of hazardous materials within the U.S.. In Canada, this responsibility falls on Transport Canada. DOT regulations are intended to promote the uniform enforcement of laws regarding the interstate movement of hazardous materials. These laws minimize the inherent dangers to life and property posed by transferring dangerous substances by common carriers engaged in interstate and foreign commerce. The regulations cover packaging, labeling, placarding, marking, filing, operator training, and emergency situations and are contained in the *Code of Federal Regulations*, Title 49.

Like the UN system, DOT has classified hazardous materials according to their primary danger and assigned standardized symbols to identify the classes. In addition to the nine classes identified in the UN system, DOT regulations cover several other types of substances (Table 7-3). Within some classes, a distinction is made between bulk and package shipments. Bulk shipments are those that equal or exceed 110 gal liquid (415 liters) or 1000 lb (450 kg) dry measure, such as tank car or tank truck. Package shipments are those that are less than 110 gal liquid or 1000 lb dry measure per package.

Table 7-2   Some Hazardous Materials by Class and
Identification Number

Hazardous Material	Hazard Class	Identification No.
Ammonium Nitrate	5	1942
Black Powder	1	—
Carbolic Acid (Phenol)	6	1671
Carbon Tetrachloride	9	1846
Gasoline	3	1203
Hydrochloric Acid	8	1789
Lauroyl Peroxide	5	2124
Oleum	8	1831
Phosphorus	4	1381
Styrene	3	2055
Uranium Hexafluoride	7	9173
Vinyl Chloride	2	1086

The following hazardous materials definitions have been abstracted from DOT regulations, 49 CFR 100-199:

Explosive

An explosive is any chemical compound, mixture, or device, the primary or common purpose of which is to function by explosion, i.e., with substantially instantaneous release of gas and heat. The three classes of explosives as well as blasting agents are defined below. It should be noted that DOT has not assigned explosives or blasting agents four-digit identification numbers.

A *Class A Explosive* will detonate or otherwise has the potential of producing maximum hazard. Class A explosive includes substances that are most likely to explode and are sensitive to shock, heat, and contamination, e.g., dynamite, nitroglycerine, blasting caps, black powder, and others.

A *Class B Explosive* functions in general by rapid combustion, or deflagration, rather than detonation and has the potential of producing flammable hazard. Class B explosive includes liquid or solid propellant explosives, some explosive devices such as special fireworks, some pyrotechnic signal devices, some smokeless powders, and others.

A *Class C Explosive* is a manufactured article containing either Class A or Class B explosive, or both, as components but in restricted quantities. It has minimum potential of producing hazard. Class C explosive includes certain types of fireworks, small arms ammunition, safety fuses, paper caps, and others.

A *Blasting Agent* is a material designed for blasting, usually consisting of a combination of fuels and oxidizers, which has been tested and found to be

Table 7-3   DOT Classification of Hazardous Materials

Class 1 - Explosives	Class A - Maximum hazard
	Class B - Flammable hazard
	Class C - Minimum hazard
	Blasting agents
Class 2 - Gases	Flammable compressed gas
	Non-liquefied compressed gas
	Liquefied compressed gas
	Compressed gas in solution
	Cryogenic liquid
Class 3 - Flammable liquids	Flammable liquid
	Combustible liquid
Class 4 - Flammable solid	Flammable solid
	Spontaneously combustible solid
	Dangerous when wet
Class 5 - Oxidizers and organic peroxides	Oxidizer
	Organic peroxide
Class 6 - Poisons	Poison A
	Poison B
	Irritating material
	Etiologic agent
Class 7 - Radioactive materials	White I label
	Yellow II label
	Yellow III label
Class 8 - Corrosive materials	
Class 9 - Other regulated materials	ORM-A
	ORM-B
	ORM-C
	ORM-D
	ORM-E

so insensitive that there is very little probability of accidental initiation to explosion or of transition from deflagration to detonation. Blasting agent includes ammonium nitrate, fuel oil, and others.

Compressed Gas

A compressed gas is any material or mixture having in the container an absolute pressure exceeding 40 psia at 70°F (276 kPa at 21°C), or a pressure exceeding 104 psia at 130°F (717 kPa at 54°C); or any flammable liquid material having a vapor pressure exceeding 40 psia at 100°F (276 kPa at 38°C). Further divisions are flammable gas, nonflammable gas, poison gas, oxygen, and chlorine.

A *Flammable Gas* is a gas that, when an ignition source is set 18 in. (45.7 cm) away from an open valve, will ignite and flash back the 18 in. to the valve (DOT test). Flammable gas includes methane, propane, hydrogen, and others.

A *Nonflammable Gas* is a gas that, when an ignition source is set 18 in. (45.7 cm) away from an open valve, may or may not ignite and will not flash back the 18 in. to the valve (DOT test). Nonflammable gas includes carbon dioxide, nitrogen, helium, and others.

A *Poison Gas (Poison A)* is an extremely dangerous poison that is either a gas or a liquid of such nature that a very small amount of the gas, or vapor of the liquid, mixed with air, is dangerous to life. Poison gas includes cyanogen, hydrocyanic acid, phosgene, and others.

Oxygen and *Chlorine* are two chemicals that have their own labels and placards; oxygen using the oxidizers label and chlorine the poison label. Both, however, indicate gases by the Class 2 marking in the lower corner of the placard.

Flammable Liquid

A flammable liquid is, generally speaking, any liquid that burns readily. The two divisions are flammable liquid and combustible liquid.

A *Flammable Liquid* is any liquid having a flash point below 100°F (38°C). Flammable liquid includes acetone, gasoline, toluene, and others. A pyrophoric liquid is any liquid that ignites spontaneously in dry or moist air at or below 130°F (54°C).

A *Combustible Liquid* is any liquid having a flash point above 100°F (38°C) and below 200°F (93°C). Combustible liquid includes fuel oil, kerosene, ethyl alcohol, and others.

Flammable Solid

A flammable solid is, generally speaking, any solid that burns rigorously and persistently. The three divisions are: flammable solid, spontaneously combustible solid, and water-reactive solid.

A *Flammable Solid* is any solid material, other than an explosive, that is liable to cause fires through friction, retained heat from manufacturing or processing, or that can be ignited readily and when ignited burns so vigorously and persistently as to create a serious transportation hazard. Flammable solid includes safety matches, sulfur, metallic calcium, and others.

A *Spontaneously Combustible Solid* is any solid material that ignites as a result of retained heat from processing, or that will oxidize to generate heat and ignite, or that absorbs moisture to generate heat and ignite. Spontaneously combustible solids include sodium hydrosulfide, and others.

A *Water-Reactive Solid* is any solid that will chemically react with water to become spontaneously flammable or to give off flammable or toxic gases in dangerous quantities. Water-reactive solid includes lithium, potassium, sodium, and others.

Oxidizer

An oxidizer is, generally speaking, a substance that yields oxygen readily to support combustion. The two divisions are oxidizer and organic peroxide.

An *Oxidizer* is any substance that yields oxygen readily to stimulate the combustion of organic matter. Oxidizer includes substances such as chlorates, permanganates, nitrates, inorganic peroxides, and others.

An *Organic Peroxide* is any organic compound containing the bivalent -O-O-structure and that may be considered a derivative of hydrogen peroxide where one or more of the hydrogen atoms have been replaced by organic radicals. Organic peroxide generally has the special property that if it is heated beyond its transportation temperatures, it will most likely deteriorate.

Poison

A poison is any chemical or substance that can kill, injure, or impair living organisms. Poison A was discussed under compressed gases. The other three divisions of poison are poison B, irritating material, and etiologic agent.

A *Poison B* is any liquid or solid substance, other than Class A poison or irritating material, that is known to be so toxic to man as to afford a health hazard during transportation, or that, in the absence of adequate data on human toxicity, are presumed to be toxic to man. Poison B includes parathion, malathion, potassium arsenate, and others.

An *Irritating Material* is any liquid or solid substance, that upon contact with fire or when exposed to air, gives off dangerous or intensely irritating fumes but not including any Poison A. Irritating material includes tear gas candles, xylyl bromide, and others.

An *Etiologic Agent* is any viable microorganism, or its toxin, that causes or may cause human disease. Etiologic agent includes anthrax, botulism, polio virus, and others.

Radioactive Material

A radioactive material is any substance, or combination of substances, that spontaneously emits ionizing radiation, and having a specific activity greater than 0.002 µCi/g. The radiation levels allowed for outside packages are limited by federal regulations. Three labels are used to identify packages containing radioactive materials:

The *Radioactive I* white label indicates that virtually no radiation exists outside the package or 0.5 mrem/hr maximum on surface, e.g., Cr^{51}, and others.

The *Radioactive II* yellow label indicates low level of radiation: 50 mrem/hr maximum on surface or 1 mrem/hr maximum at 3 ft, e.g., I^{131}, and others.

The *Radioactive III* yellow label indicates increased level of radiation: 200 mrem/hr maximum on surface or 10 mrem/hr maximum at 3 ft, e.g., Sr^{90}, and others.

Corrosive Material

A corrosive material is any liquid or solid that causes visible destruction or irreversible alterations in human skin tissue at the site of contact, or a liquid that has a severe corrosion rate on steel. A material is considered to be corrosive if, when tested on the intact skin of the albino rabbit, the structure of the tissue at the site of contact is destroyed or changed irreversibly after an exposure period of 4 hr or less. A liquid is considered to be corrosive if its corrosion rate exceeds 0.250 in. (6.350 mm) per year on steel at a test temperature of 130°F (54°C). Corrosive material includes ammonium hydroxide, benzoyl chloride, nitric acid, sulfuric acid, and others.

Other Regulated Material

Other regulated material (ORM) is any material that does not meet the definition of a hazardous material; or any material that possesses one or more of the characteristics described in ORM-A through ORM-E below.

ORM-A is any material that has an anesthetic, irritating, noxious, toxic, or other similar property and that can cause extreme annoyance or discomfort to passengers and crew in the event of leakage during transportation. ORM-A includes dry ice, acetaldehyde, bone oil, and others.

ORM-B is any material capable of causing significant damage to a transport vehicle from leakage during transportation. ORM-B includes metallic mercury, copper chloride, barium oxide, and others.

ORM-C is any material that contains other inherent characteristics not described under ORM-A or ORM-B and is unsuitable for shipment unless properly identified and prepared for transportation. ORM-C includes excelsior, wet feed, fishcrap, and others.

ORM-D is any material such as consumer commodity that presents a limited hazard during transportation due to its form, quantity, and packaging. ORM-D includes hair spray, shaving cream, and others.

ORM-E is any material that is not included in any hazard class yet subject to the requirements of this subclass because it is a hazardous waste or hazardous substance. ORM-E includes ferrous sulfate, ferric fluoride, potassium chromate, and others.

THE DOT PLACARDING/LABELING SYSTEM

The Department of Transportation (DOT) placards are 10 3/4 in. (27.3 cm) square, on-point, diamond-shaped signs that are fixed to each side and to each end of any motor vehicle or rail car containing specified amounts of hazardous materials. Table 7-4 summarizes the possible hazards indicated by a placard and the quantity of hazardous materials that may be present. The DOT placard indicates the primary or most dangerous hazard of the material and may display a specific commodity number on the placard or on an accompanying orange panel. The placards have distinctive colors for each of the major divisions so that the general type of hazard can be identified from a distance. The specific details of placard and label use are contained in 49 CFR 100-177.

A "DANGEROUS" placard may be used for mixed loads of two or more classes of hazardous materials. The use of this generic placard is accepted in place of the separate placards specified for each class. The "DANGEROUS" placard may not be used, however, when 5000 lb (2268 kg) or more of one class of hazardous material is loaded at one facility. The specific placard for each class must then be used. Since there is no "EXPLOSIVE C" or "IRRITANT" placard, the "DANGEROUS" placard is used. It is also important to note that transportation of any quantity of *Explosive A, Explosive B, Poison Gas, Flammable Solid, and Radioactive III,* must be specifically placarded.

Any person who offers a hazardous material for transportation must label the package. DOT labels look like small placards. They are 4 in. (10.2 cm) squares. When their use is required by regulations, labels must be printed on or affixed to the surface of the package near the proper shipping name. Even when their use is not required, labels may be affixed to packages provided each one of them represents a hazard of the material in the package. If two or more different labels are used, they are to be displayed next to each other.

THE NFPA HAZARD IDENTIFICATION SYSTEM

The National Fire Protection Association (NFPA) 704 system is a nationally recognized method of identifying the presence of hazardous materials at commercial, manufacturing, institutional, and other fixed storage facilities. The NFPA 704 system uses a standard diamond symbol with color backgrounds and numerical ratings. The color identifies the kind of hazard and the number identifies the degree of hazard. The diamond is divided into four smaller diamonds with the background color indicating the specific areas as shown in Figure 7-1. The left is always blue and refers only to *health*. The top is always red and refers only to *flammability*. The right is always yellow and refers only to *reactivity*. The bottom is always white and is reserved for identifying those materials that may cause special problems or require special fire-fighting techniques. Each diamond, except the white, has a number between 0 and 4 to indicate the degree of hazard.

Table 7-4   DOT Placarding System

Placard	Hazardous Materials That May Be Present
Explosive A	ANY QUANTITY - Explosive A
	ANY QUANTITY - Explosive A and B loaded together
Explosive B	ANY QUANTITY - Explosive B
Blasting Agent	1000 lb or more - Blasting agents
Poison Gas	ANY QUANTITY - A poison gas
Flammable Gas	1000 lb or more - A flammable gas
Nonflammable Gas	1000 lb or more - A nonflammable gas
	1000 lb or more - Gaseous oxygen
	110 gal or less - Liquefied chlorine
Chlorine	110 gal or more - Liquefied chlorine
Oxygen	1000 lb or more - Liquefied oxygen
Flammable	1000 lb or more - A flammable liquid
	1000 lb or more - A flammable solid
Combustible	110 gal or more - A combustible liquid
Flammable Solid	1000 lb or more - A flammable solid
Flammable Solid W	ANY QUANTITY - A flammable solid which is DANGEROUS WHEN WET
Oxidizer	1000 lb or more - An oxidizer
Organic Peroxide	1000 lb or more - An organic peroxide
Poison	1000 lb or more - A class B poison or fluorine
Radioactive	ANY QUANTITY - Radioactive material with a radioactive class III label
Corrosive	1000 lb or more - A corrosive material
Dangerous	1000 lb or more - Materials with Explosive C labels
	1000 lb or more - A combination of materials
	1000 lb or more - An irritating material

Health

In general, health hazard in fire-fighting is that of a single exposure that may vary from a few seconds up to an hour. The physical exertion demanded in fire-fighting or other emergency conditions may be expected to intensify the effects of any exposure. Only hazards arising out of an inherent property of the material are considered. The following explanation is based upon protective equipment normally used by fire-fighters.

Degree 4 - Indicates materials which will endanger the health of the exposed fire-fighters. The inhalation of a few whiffs of the vapor could cause death, or the vapor or liquid could cause a fatal penetration of the normal structural fire-fighting clothing and breathing apparatus available to the average fire department. Or, ordinary structural fire-fighting clothing and breathing apparatus

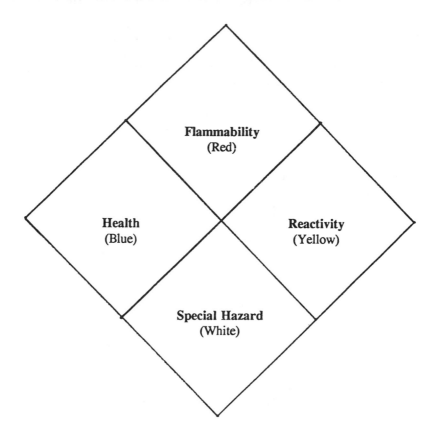

Figure 7-1   NFPA diamond symbol.

cannot provide adequate protection against skin contact or inhalation with these materials.

Degree 3 - Indicates materials which are extremely hazardous to health, yet these areas may be entered with extreme caution. Encapsulating chemical-resistant protective clothing, including self-contained breathing apparatus, gloves and boots must be worn. No skin surface should be exposed.

Degree 2 - Indicates materials which are hazardous to health, yet these areas may be entered freely with full-faced mask self-contained breathing apparatus.

Degree 1 - Indicates materials which are only slightly hazardous to health. It may be desirable to wear self-contained breathing apparatus.

Degree 0 - Indicates materials which on exposure under fire conditions would offer no hazard beyond that of ordinary combustible material.

Flammability

Susceptibility to burning is the basis for assigning degrees within this category. The method of attacking the fire is influenced by this susceptibility factor.

Degree 4 - Indicates extremely flammable gases or extremely volatile flammable liquids. The flow of such flammable gases or liquids is to be shut off when fighting the fire and cooling water streams are to be kept on exposed tanks or containers.

Degree 3 - Indicates materials which can be ignited under almost all normal temperature conditions. Water may not be an effective agent of fighting the fire because of the low flash point.

Degree 2 - Indicates materials which must be moderately heated before ignition will occur. Water spray may be used to extinguish the fire because the material can be cooled below its flash point.

Degree 1 - Indicates materials which must be preheated before ignition can occur. Water may cause frothing if it gets below the surface of the liquid and turns to steam. However, a water fog gently applied to the surface will cause a frothing that will extinguish the fire.

Degree 0 - Indicates materials which will not burn.

Reactivity

The assignment of degrees in the reactivity category is based upon the susceptibility of materials to release energy either alone or in combination with water. Fire exposure, along with conditions of shock and pressure, are factors to be considered.

Degree 4 - Indicates materials which are capable of self-detonation, self-explosive decomposition, or self-explosive reaction at normal temperatures and pressures. It includes materials sensitive to mechanical or localized thermal shock. If a chemical with this hazard rating is in an advanced or massive fire, the area must be evacuated.

Degree 3 - Indicates materials which are capable of self-detonation, self-explosive decomposition, or self-explosive reaction, but they require a strong initiating source or they must be heated under confinement before ignition. It includes materials sensitive to thermal or mechanical shock at elevated temperatures and pressures or those that react explosively with water without requiring heat or confinement. Fire-fighting should be conducted from an explosion-resistant location.

Degree 2 - Indicates materials which are normally unstable and readily undergo violent chemical change but they do not detonate. It includes materials that can undergo chemical change with rapid release of energy at normal temperatures and pressures or that can undergo violent chemical change at elevated temperatures and pressures. In addition, it includes those materials that may react violently with water or that may form potentially explosive mixtures with water. In advanced or massive fire, fire-fighting should be conducted from a safe distance or from a protected location.

Degree 1 - Indicates materials which are normally stable but they may become unstable at elevated temperatures and pressures or they may react with water with a nonviolent release of energy. Caution must be used in approaching the fire and applying water.

Degree 0 - Indicates materials which are normally stable even under fire exposure conditions and they are not reactive with water. Normal fire-fighting procedures may be used.

Special Hazards

The bottom diamond uses a symbol to indicate water reactivity, oxidation capability, radioactivity, or any hazard. For example, materials that in contact with water demonstrate unusual reactivity are identified with the letter W with a horizontal line through the center. Materials that are oxidizers are identified by the letters OX. Materials that possess radioactivity hazards are identified with the standard radioactivity symbol. Since 1957, the NFPA Committee on Identification of Fire Hazards of Materials has done considerable work on this system and the classification of hazardous materials. However, it should be pointed out that the system assumes that only short-term exposure will be permitted.

Laboratory workers must know that they cannot depend exclusively on placard, label, sign, or symbol identification systems to ensure their safety. These systems are only as reliable as the people who place and maintain them. The possibility of human error is ever present.

8 CHEMICAL HANDLING PROCEDURES

Workers in scientific laboratories are exposed to numerous hazards. Many of these hazards are generally well known—while other hazards, especially those of a chemical nature, are yet to be discovered. To compound the risks of handling hazardous chemicals, workers in research laboratories are relentlessly exploring the unknown and conquering new horizons. In contrast to research laboratories, most other laboratories usually handle only small amounts of materials and exposure to a particular chemical seldom extends over a protracted time. In this respect, industrial laboratories differ little from university, governmental, or other organizational laboratories. As a result, workers in research laboratories are constantly faced with the potential existence of undiscovered as well as undisclosed hazards.

A chemical hygiene plan should contain, at a minimum, the following standard operating procedures. The procedures to be discussed in this chapter are examples of good laboratory practice. Each laboratory is different and local conditions will govern the utility of the plethora of hazard information. The reader must alter, change, modify, adapt, and extend the concept of the specific item in the development and preparation of a chemical hygiene plan.

In order to prepare a comprehensive chemical hygiene plan, a thorough inventory of all hazardous chemicals must first be conducted and the inventory reported to the chemical hygiene officer. The inventory must include both the quantity and location of each chemical. Chemicals which may be considered to be innocuous, but in fact are hazardous, must also be included. A list of the hazards and the appropriate precautionary measures can then be developed from the inventory information. Needless to say, the inventory must be accurate and it must be updated whenever warranted and the chemical hygiene plan is to be revised accordingly.

HANDLING PROCEDURES FOR FLAMMABLE CHEMICALS

The obvious hazard of flammable or fire-hazard chemicals is the damage caused to life and property when they burn. Since a large percentage of fires are fueled by liquids, it is important to recognize that it is the vapor of liquids that actually burn. It is also important from the safety point of view to note that, with a few exceptions, flammable liquid vapors are more dense than air. Therefore, at room temperature, most vapors tend to settle to the lowest possible point.

Physical Properties

The prevention of unwanted fires and vapor or gas explosions requires a knowledge of the flammability characteristics of combustible vapor and gases. The flammability characteristics of a material in turn depends on such properties as its flash point, fire point, boiling point, autoignition temperature, and flammability range. Possession of this information is vital for prevention of fires.

Flash Point. This is the minimum temperature at which a liquid fuel gives off sufficient vapors to form an ignitable mixture with air near the surface. At this temperature, the ignited vapor will flash, but will not continue to burn. It is important to know that flammable liquids themselves will not burn but the vapors they produce can ignite with explosive force. The higher the temperature of the liquid, the more vapors are emitted. Flammable liquids such as gasoline and acetone have flash points well below 100°F (38°C). Combustible liquids such as fuel and lubricating oils have flash points above 100°F (38°C). Since there exist tens of thousands of liquids that fit into these two classes, it stands to reason to say that there also exists a wide range of flash points within each class. The National Fire Protection Association (NFPA) has subdivided each class, flammable and combustible, into more definitive classes so that safer storage of these materials can be achieved. Table 8-1 summarizes the NFPA scheme. Flammable gases have no flash points since they are already in the gaseous state. With the exception of a few solids, e.g., naphthalene with a flash point of about 174°F (79°C), solids are not considered to have flash points either.

Fire Point. This is the temperature at which a liquid fuel produces enough vapors to support combustion once ignited. For most substances, the fire point is just a few degrees above the flash point.

Autoignition Temperature. This is the minimum temperature to which a flammable or combustible liquid must be heated to initiate self-sustained combustion without an ignition source. All flammable or combustible liquids have autoignition temperatures which are always higher than the flash points and boiling points of the corresponding liquids.

Flammability Range. This is the percentage of the vapor or gas in air that will burn if ignited. Below the lower flammable limit (LFL), a vapor or gas is too lean to burn. Above the upper flammable limit (UFL), the vapor or gas is too rich to

**Table 8-1   NFPA Classification of Flammable and
Combustible Liquids**

Classification	Flash Point and Boiling Point
Class IA flammable liquid	Flash point below 73°F (22.8°C) Boiling point below 100°F (37.8°C)
Class IB flammable liquid	Flash point below 73°F (22.8°C) Boiling point at or above 100°F (37.8°C)
Class IC flammable liquid	Flash point at or above 73°F (22.8°C) and below 100°F (37.8°C)
Class II combustible liquid	Flash point at or above 100°F (37.8°C) and below 140°F (60°C)
Class IIIA combustible liquid	Flash point at or above 140°F (60°C) and below 200°F (93.4°C)
Class IIIB combustible liquid	Flash point at or above 200°F (93.4°C)

burn. Within the upper and lower flammability limits, also referred to as explosive limits, the vapor or gas will burn rapidly if ignited. Table 8-2 displays the properties of some common flammable liquids and Figure 8-1 displays the relationships of flash point, boiling point, autoignition temperature, and flammability range of a vapor–air mixture of a flammable liquid.

OSHA Definitions

According to the OSHA Laboratory Standard, a flammable chemical is defined as a substance that falls into one the following four categories:

A *flammable aerosol* is an aerosol that, when tested by the method described in 16 CFR 1500.45, yields a flame projection exceeding 18 in. (45.7 cm) at full valve opening, or a flashback (a flame extending back to the valve) at any degree of the valve opening.

A *flammable gas* is a gas that, at ambient temperature and pressure, forms either a flammable mixture with air at a concentration of 13% by volume or less or a range of flammable mixtures with air wider than 12% by volume, regardless of the lower limit.

A *flammable liquid* is any liquid having a flash point below 100°F (37.8°C) except any mixture having components with flash points of 100°F (37.8°C) or higher, the total of which make up 99% or more of the total volume of the mixture.

A *flammable solid* is any solid, other than a blasting agent or explosive, as defined in 29 CFR 1910.109(a), that is liable to cause fire through friction, absorption of moisture, or retained heat from manufacturing or processing, or that can be ignited readily and when ignited burns so vigorously and persistently

Table 8-2   Properties of Some Common Flammable Liquids

Chemical	Flash Point (°F)	Flammability Limits		NFPA Rating
		LFL (%)	UFL (%)	
Acetaldehyde	−38	4.0	60	2-4-2
Acetic Acid (Glacial)	103	4.0	20	2-2-1
Acetone	−4	2.2	13	1-3-0
Acetonitrile	42	3.0	16	2-3-1
Aniline	158	1.3	11	3-2-0
Benzene	12	1.3	7	2-3-0
Carbon Disulfide	−22	1.3	50	2-3-0
Cyclohexane	−4	1.3	8	1-3-0
Diethyl Ether	−49	1.9	36	2-4-1
Dimethylformamide	136	2.2	15	1-2-0
Ethyl Acetate	24	2.0	12	1-3-0
Ethyl Alcohol	55	3.3	19	0-3-0
Ethylene Oxide	<0	3.6	100	2-4-3
Gasoline	-45	1.4	8	1-3-0
n-Hexane	−7	1.1	8	1-3-0
Hydrazine (Anhydrous)	100	2.9	98	3-3-2
Methyl Alcohol	52	6.0	36	1-3-0
Methyl Ethyl Ketone	16	1.7	11	1-3-0
n-Pentane	−40	1.5	8	1-4-0
Propyl Acetate	55	1.7	8	1-3-0
Propylene Oxide	−35	2.8	37	2-4-2
Pyridine	68	1.8	12	2-3-0
Styrene	90	1.1	6	2-3-2
Tetrahydrofuran	6	2.0	12	2-3-0
Toluene	40	1.2	7	2-3-0
Xylene	81	1.1	7	2-3-0

as to create a serious hazard. A chemical shall be considered to be a flammable solid if, when tested by the method described in 16 CFR 1500.44, it ignites and burns with a self-sustained flame at a rate greater than one-tenth of an inch (2.54 mm) per second along its major axis.

In addition to the four categories of flammable chemicals defined above, an explosive chemical is defined in the OSHA Laboratory Standard as a substance that causes a sudden, almost instantaneous release of pressure, gas, and heat when subjected to sudden shock, pressure, or high temperature.

OSHA also considers all chemicals with a flash point below 200°F (93.3°C) to be *fire-hazard* chemicals. Fire-hazard chemicals are to be used only in vented hoods and away from sources of ignition. When not in use, fire-hazard chemicals must be stored in a flammable solvent storage area or in storage cabinets designed for flammable materials.

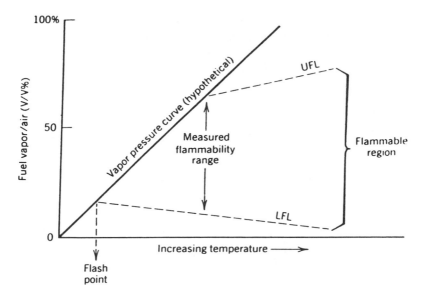

Figure 8-1   Flammability factors of a flammable liquid.

In all work with flammable or fire-hazard chemicals, the OSHA requirement in 29 CFR, subpart H and L; NFPA Manual 30, *Flammable and Combustible Liquid Codes;* and NFPA Manual 45, *Fire Protection for Laboratories
Using Chemicals,* must be followed.

Handling Procedures

Among the most hazardous liquids are those with flash points at room
temperature or lower, particularly if their range of flammability is broad. Materials
having flash points higher than the maximum ambient summer temperature do not
ordinarily form ignitable mixtures with air under normal conditions but as shown
in Table 8-2, many commonly used substances are potentially very hazardous,
even under relatively cool conditions. The following precautions must always be
followed in the safe handling of flammable chemicals:

(1)   For a fire to occur, three distinct conditions must exist simultaneously:
      a concentration of flammable gas or vapor that is within the flammable
      limits of the substance; an oxidizing atmosphere, usually air; and a
      source of ignition. Removal of any of the three will prevent the start
      of a fire or extinguish an existing fire. In most situations, air cannot be
      excluded. The problem, therefore, is usually resolved through the
      prevention of the coexistence of flammable vapors and an ignition
      source. Since the probability of spillage of a flammable liquid is
      always a possibility, strict control of ignition sources is mandatory.

(2) Various sources, e.g., electrical equipment, open flames, static electricity, burning tobacco, lighted matches, and hot surfaces, can cause ignition of flammable substances. When these materials are used in the laboratory, close attention should be given to all potential sources of ignition in the vicinity.

(3) The vapor of all flammable liquids is heavier than air and capable of traveling considerable distances. This possibility must be recognized and special note taken of ignition sources at a lower level than that at which the substance is being used. Flammable vapors from spillage have been known to descend into stairwells and elevator shafts and ignite on a lower story. If the path of vapor within the flammable range is continuous, the flame will propagate itself from the point of ignition back to its source.

(4) Metal lines and vessels discharging flammable liquids must be properly bonded and grounded to discharge static electricity. When nonmetallic containers are used, the bonding can be made to the liquid rather than to the container. If no solution to the static problem can be found, then all processes must be carried out as slowly as possible to give the accumulated charge time to disperse.

(5) Ventilation is one of the most effective ways to prevent the formation of flammable mixtures. An exhaust hood must be used whenever appreciable quantities of a flammable liquid are transferred from one container to another, allowed to stand in open containers, heated in open containers, or handled in any other way.

(6) Leakage or escape of flammable gases can produce an explosive atmosphere in the laboratory. Acetylene, hydrogen, ammonia, hydrogen sulfide, and carbon monoxide are especially hazardous. Acetylene and hydrogen have very wide flammability limits, which adds greatly to their potential fire and explosion hazard.

HANDLING PROCEDURES FOR PYROPHORIC CHEMICALS

Pyrophoric chemicals are those chemicals that rapidly undergo air oxidation or react with moisture and spontaneously ignite in air. Thus, such chemicals do not require an external ignition source. Ignition of a pyrophoric chemical may be delayed or only occur if the chemical is finely divided or spread as a diffuse layer, e.g., titanium powder, and mixtures of tributyl phosphine isomers, respectively. On the other hand, ignition may be virtually instantaneous, the delay time being measured in milliseconds, as is seen in trimethylaluminum. Pyrophoricity is sometimes put to practical application, as in the use of ferrocerium for lighter flints, and injection of trimethylaluminum to reignite jet engines operating at high altitudes. Examples of some common pyrophoric chemicals are given in Table 8-3.

Table 8-3   Some Common Pyrophoric Materials

Classification	Examples
Finely divided metals	Calcium, titanium, zirconium
Metal hydrides or	Germane
nonmetal hydrides	Diborane
Metal carbonyls	Octacarbonyldicobalt
	Pentacarbonyliron
Metal alkyls or	Trimethylaluminum
nonmetal alkyls	Trimethylphosphine
Alkylated metal alkoxides or	Diethylethoxyaluminum
nonmetal halides	Dichloromethylsilane
Used hydrogenation catalysts	Lithium aluminum hydride

All pyrophoric chemicals should be stored in tightly closed containers and when being transferred they should be kept under an inert atmosphere or liquid. Wherever such chemicals are being used in an operation, use of an inert atmosphere and specialized handling techniques are essential for safety. Again, failure to properly store and handle these chemicals can lead to severe and unexpected fires.

HANDLING PROCEDURES FOR HYPERGOLIC MIXTURES

Another group of chemicals that can also lead to unexpected fires are hypergolic mixtures. The mixture of such pairs of chemicals can create enough heat of reaction to cause ignition. Again, an ignition source is not needed. The chemical reaction range of hypergolic mixtures varies from slow and barely visible to an instantaneous and highly visible explosive force.

Examples of such mixtures are numerous. Perchloric acid and magnesium powder produce a white-hot flame within microseconds of mixing. The same is true of nitric acid and phenol. Acetone and 85% nitric acid constitutes a hypergolic mixture, as does concentrated nitric acid and triethylamine. Red fuming nitric acid forms such a mixture with many different aromatic amines. Divinyl ether ignites after one millisecond when mixed with 96% nitric acid that contains sulfuric acid at concentrations at less than 5%. A 90% solution of potassium permanganate in red fuming nitric acid will ignite when mixed with alcohols. Ketones, ethers, and liquid alcohols form hypergolic mixtures with solid potassium permanganate. Liquid fluorine is particularly dangerous. Nitrogen tetroxide reacts explosively with gasoline or other flammable liquids. Fluorine reacts with most elements. Mixed with hydrogen, or substances that contain hydrogen, fluorine is self igniting.

Fire prevention involves more than just avoiding ignition sources in the handling of hypergolic mixtures. The handling must ensure that pairs of hypergolic materials are not accidentally mixed. The short time interval between mixing of hypergolic materials and ignition justifies expending time prior to mixing unknown chemicals to discover if indeed an operator is working with such chemicals.

HANDLING PROCEDURES FOR REACTIVE/UNSTABLE CHEMICALS

The OSHA Laboratory Standard defines a reactive or unstable chemical as a substance that, in the pure state, or as produced or transported, will vigorously polymerize, decompose, or condense, and will become self-reactive under conditions of shocks, pressure, or temperature. A chemical is considered *water-reactive* if it reacts with water to release a gas which is either flammable or presents a health hazard.

Due to advances in modern technology, more and more reactive and unstable materials are being used for various processes. Used under controlled conditions, reactive and unstable materials can be useful and beneficial. Unfortunately, reactive or unstable chemicals under chemical emergency conditions can be extremely destructive to life and property.

A chemical is unreactive or stable if its rate of change is relatively slow. A chemical is reactive if its rate of reaction is readily observable. The word "routine" usually implies a "safe" reaction when it is used to describe a chemical reaction. A reaction is considered safe if its reaction rate is relatively slow. For a reaction to begin, the energy of activation must first be supplied to the reacting chemicals so that the necessary transition state can be attained in order for the reaction to proceed. Once the reaction starts, in most cases energy is released in the form of heat, and unless this is removed at the same rate it tends to increase the temperature of the reaction system. Such reactions are described as exothermic reactions, and most reactions are of this type. The other type of reaction occurs when energy is absorbed into the structure of the reaction products rather than being released. Such reactions are described as endothermic reactions. In endothermic reactions the absorption of heat energy tends to reduce the temperature, and heat energy usually has to be supplied to the reaction system to keep the reaction going. Due to their relatively high bond energy content, endothermic compounds tend to be less stable than the products of exothermic reactions.

The rate of reaction almost always increases as the temperature increases. If the heat evolved in a reaction is not dissipated, the reaction rate can increase until an explosion results. Consider this factor particularly when scaling up experiments to ensure that sufficient cooling and surface for heat exchange can be provided.

To readily understand the subject of reactivity, which is defined as the ability of a chemical to undergo a chemical reaction, it would be desirable to draw some distinction between chemical structures: those that exhibit instability in the absence of contact with appreciable amounts of other chemicals; and those that exhibit a high level of reactivity in combination with other chemicals or materials, including atmospheric oxygen and moisture.

Some chemicals decompose when heated. Slow decomposition is often unnoticeable on a small scale, but on a large scale or if the evolved heat and gases are confined, an explosion can occur. The heat-initiated decomposition of some substances, such as certain peroxides, is almost instantaneous. Light, mechanical shock, and certain catalysts are additional initiators of explosive reactions. For example, hydrogen and chlorine react explosively in the presence of ultraviolet light. Examples of shock-sensitive chemicals include acetylides, azides, fulminates, nitro compounds, organic nitrates, picric acid, and many peroxides. Many substances catalyze the explosive polymerization of acrolein, and many metal ions can catalyze the violent decomposition of hydrogen peroxide.

Structure and Reactivity

Detailed recognition of which structural features will confer suitable levels of reactivity upon potential reaction components is the art of the synthetic chemist. It is possible, however, to indicate some general structural features often associated with high levels of reactivity in single chemicals and these are listed in Table 8-4. Many of the these compounds in Table 8-4 contain a high level of multiple bondage and are endothermic compounds. Some also possess very low values of energy of activation. These compounds are highly hazardous because they will decompose with evolution of heat. The accumulated heat will rapidly accelerate the rate of decomposition, often to the point of explosion. It may be broadly anticipated that chemical compounds with similar structures will show greatly reduced stability. In general, the smaller the molecular weight, the larger the energy of decomposition per unit weight of the chemical compound. Therefore, small molecules are potentially more hazardous than larger ones of the same structural type.

Organic Peroxides

Organic peroxides are a special class of compounds with unusual stability problems which make them among the most hazardous chemicals normally handled in laboratories. As a class, organic peroxides are low-power explosives. The hazard exists in their extreme sensitivity to shock, sparks, or other forms of accidental ignition. A number of organic peroxides are more sensitive to shock than most primary explosives such as trinitrotoluene.

Organic peroxides have a specific half-life, or rate of decomposition, under any given set of conditions. A low rate of decomposition may auto-accelerate and

Table 8-4   Some Common Reactive Compounds

Classification	Examples
Acetylenic compounds	Acetylene
	Copper(I) acetylide
	Ethoxyacetylene
Azides	Benzenesulfonyl azide
	Lead(II) azide
	Silver azide
Azo compounds	Azomethane
	Diazomethane
Chloro/perchloro compounds	Lead perchlorate
	Potassium chlorite
	Silver chlorate
Fulminates	Copper(II) fulminate
	Mercury(II) fulminate
	Silver fulminate
Nitro compounds	Nitroglycerine
	Nitromethane
	Trinitrotoluene
Nitrogen-containing compounds	Silver amide
	Silver nitride
Picrates	Picric acid (dry)
	Lead picrate
Peroxides	Diacetyl peroxide
	Zinc peroxide
Strained ring compounds	Benzvalene
	Prismane
Polymerizable compounds	Butadiene
	Styrene
	Vinyl chloride

cause a violent explosion, especially in bulk quantities. In addition, organic peroxides are sensitive to heat, friction, impact, and light, as well as to strong oxidizing and reducing agents. All organic peroxides are highly flammable, and fires involving bulk quantities of peroxides must be approached with extreme caution. A peroxide present as a contaminant in a reagent or solvent can change the course of a planned reaction.

Organic peroxides are frequently formed due to the inherent instability of a certain chemical structure. Such peroxidizable chemicals invariably contain the structure of an auto-oxidizable hydrogen atom which is activated by adjacent structural features and/or the presence of actinic radiation, so that it reacts slowly under ambient conditions with atmospheric molecular oxygen, initially to form a hydroperoxide:

$$RH + O_2 \longrightarrow ROOH$$

In itself this may be highly unstable — though usually a further, also slow, reaction step involving addition, rearrangement, or disproportionation of the initially formed hydroperoxide is necessary to develop potentially hazardous peroxidic instability, which may only become apparent after heating or concentration by evaporation. Three classes of peroxidizable chemicals have been identified:

Class A peroxidizable chemicals include those chemicals which readily form explosive peroxides without concentration. Many of these peroxides separate from solution. All Class A peroxidizable chemicals have caused fatalities. Examples of Class A peroxidizable chemicals include: diisopropyl ether, divinyl acetylene, vinylidene chloride, and others. These chemicals are recommended for testing for presence of peroxides at maximum of 3-month intervals after opening. If the test is positive, proper disposal is a necessity. Two additional chemicals included in the Class A peroxidizable chemical classification are potassium metal and sodium amide. Both, in contact with air, forms the superoxide rather than a true peroxide.

Class B peroxidizable chemicals include those chemicals that peroxidize but only become hazardous on evaporative concentration. Many of the ether solvents are included in this category, e.g., diethyl ether, divinyl ether, dioxane, ethylene glycol dimethyl ether, tetrahydrofuran, and others. Other examples of Class B peroxidizable chemicals are various unsaturated hydrocarbons, e.g., acetal, cumene, cyclohexene, diacetylene, methylcyclopentane, methyl acetylene, methyl isobutyl ketone, tetrahydronaphthalene, and others. Each of these chemicals is recommended for testing for presence of peroxides at maximum of 12 months after opening, and again if the test is positive, for proper disposal.

Class C contains those peroxidizable chemicals that can also polymerize exothermically when initiated by the peroxide content. These monomers include acrylic acid, acrylonitrile, butadiene, chloroprene, chlorotrifluoroethylene, methyl methacrylate, styrene, tetrafluoroethylene, vinyl acetate, vinyl acetylene, vinyl chloride, vinyl pyridine, and others. Testing and disposal requirements for these chemicals are the same as for Class B.

Handling Procedures

The general precautions for handling these three classes of peroxidizable chemicals are summarized below:

(1) All peroxidizable chemicals must be prevented from peroxide formation as far as possible. Preventive measures include storage in

opaque containers and exclusion of air, preferably by nitrogen atmosphere, except for several Class C monomers that contain inhibitors for storage and which need limited access of air for effective inhibition.

(2)   All peroxidizable chemicals must be tested for peroxide presence before any heating or evaporation is attempted.

(3)   The sensitivity of most peroxides to shock and heat can be reduced by dilution with inert solvents, such as aliphatic hydrocarbons. However, toluene is known to induce the decomposition of diacyl peroxides.

(4)   Metal spatulas must not be used to handle peroxides because contamination by metals can lead to explosive decomposition. Ceramic or wooden spatulas may be used.

(5)   Friction, grinding, and all forms of impact must be avoided near peroxides, especially solid peroxides. Glass containers with screw cap lids or glass stoppers are not to be used. Polyethylene bottles with screw-cap lids may be used.

(6)   To minimize the rate of decomposition, peroxides must be stored at the lowest possible temperature consistent with their solubility or freezing point. Liquid or solutions of peroxides are not to be stored at or lower than the temperature at which the peroxide freezes or precipitates because peroxides in these forms are extremely sensitive to shock and heat.

(7)   Pure peroxides must never be disposed of directly. Peroxides must be diluted before disposal. Dilute small quantities of 25g or less of peroxides with water to a concentration of 2% or less. Transfer the solution to a polyethylene bottle containing an aqueous solution of a reducing agent such as ferrous sulfate or sodium bisulfite. The material can then be handled like any other waste chemical. However, it must not be mixed with other waste chemicals for disposal.

(8)   Any peroxidizable chemical in containers showing corrosion or damaged or suspect closures must be immediately disposed of with great care and without being opened.

HANDLING PROCEDURES FOR WATER-REACTIVE CHEMICALS

The most common substance, other than air, likely to come into contact with chemicals is water. Water-reactive materials react chemically on contact with water. The chemical reactions of water-reactive materials vary. Lithium and finely divided magnesium burn with such intensity that they decompose water into separate hydrogen and oxygen molecules. The hydrogen then burns while the oxygen supports the combustion. Adding water to these reactions intensifies the fire rather than suppressing it. Nonburning magnesium powder, potassium, and rubidium decompose water. The reaction may create enough heat to ignite the hydrogen. Sodium and cesium react explosively on contact with water. Other water-reactive materials produce highly flammable gases when in contact with water, e.g., acetylene gas is formed when water contacts calcium carbide. Other chemicals which may react exothermically and violently with water, particularly if in limited amounts, are:

(1) Alkali and alkaline earth metals, e.g., sodium, magnesium, and others.
(2) Anhydrous metal halides, e.g., aluminum bromide, germanium chloride, and others.
(3) Anhydrous metal oxides, e.g., cesium trioxide, calcium oxide, and others.
(4) Nonmetal halides, e.g., boron tribromide, phosphorus pentachloride, and others.
(5) Nonmetal oxides, e.g., sulfur trioxide, and others.
(6) Nonmetal halide oxides, e.g., phosphoryl chloride, sulfinyl chloride, and others.
(7) Organometallics, e.g., aluminum alkyls, and others.

Handling Procedures

In the handling of water-reactive chemicals, obviously utmost care must be taken to avoid the contact of such materials with water. Special storage facilities must be constructed to prevent accidental contact with water. Prevention of contact of water-reactive chemicals is best accomplished by eliminating all sources of water in the storage area, e.g., areas where large quantities of water-reactive chemicals are stored should not have automatic sprinkler systems. Storage facilities for such chemicals should be of fire-resistant construction, and other combustible materials should not be stored in the same area.

HANDLING PROCEDURES FOR CORROSIVE CHEMICALS

Corrosive chemicals cause local reactions in humans at the site of chemical contact. The sites of contact are typically the skin and eyes, but this is not always true. For example, accidental ingestion of a corrosive chemical by mouth could have local effects inside the mouth, in the esophagus, the stomach, or even the intestines. Local effects, usually after a single exposure, range from minor irritation to severe tissue destruction. Some local effects may require repeated exposures to develop. Other corrosive chemicals can burn a hole through the skin down to the bone and beyond in a single exposure. Generally, corrosive chemicals of even a low concentration and dosage can destroy the tissues of the eye or the lining of the stomach or lungs.

It is well known that liquid chemicals are capable of causing corrosive reactions through either spillage or splashing. It must be further realized that gases, vapors, fumes, and solids are also capable of causing corrosive reactions. Whereas liquid chemical corrosivity is almost instantaneous, solid chemical corrosivity is somewhat slower and, therefore, quick removal of the solid chemical can limit the extent of the damage. However, there is an additional danger that solid chemicals often pose. That danger is the heat given off when solid chemicals come in contact with moisture on the tissue. The heat generated can be so great as to cause thermal burns and tissue destruction in addition to the corrosivity of the chemical. Since liquid and solid chemicals are visible but their gases and vapors are generally invisible, they can be an insidious danger. Corrosive gases, vapors, fumes, mists, and dusts can destroy the lining of the lungs upon inhalation resulting in death, or in a chemical pneumonia that leaves the victim not only ill but disabled.

Classes of Corrosive Chemicals

The major classes of corrosive chemicals are strong acids and bases, dehydrating agents, and oxidizing agents. Some chemicals, e.g., nitric acid, belong to more than one classification. Nitric acid is classified both as a corrosive chemical and as an oxidizer.

Strong Acids. All concentrated strong acids can damage the skin and mucous membrane. Exposed areas must be flushed promptly with copious amounts of water. Nitric, chromic, and hydrofluoric acids are particularly damaging due to the severe burns they inflict. Hydrofluoric acid produces slow-healing and painful burns. It should be used only after thorough familiarization with its properties and safe handling procedures.

Strong Bases. The common strong bases are potassium hydroxide, sodium hydroxide, and ammonia. Ammonia is a severe irritant and should be used only in a well-ventilated area. The alkali metal hydroxides are extremely damaging to the eye and mucous membranes. If exposure to alkali metal hydroxides occurs, the affected areas should be washed at once with copious amounts of water. In addition, an ophthalmologist must be contacted to evaluate the need for further treatment.

Dehydrating Agents. The corrosive dehydrating agents include concentrated sulfuric acid, sodium hydroxide, phosphorus pentoxide, and calcium oxide. Extreme heat results from mixing dehydrating agents with water. The proper mixing of dehydrating agent with water is accomplished by adding the agent to water to avoid violent reaction and spattering. Due to the affinity of dehydrating agents for water, these substances cause severe burns on contact with the skin. Affected areas must be washed promptly with copious amounts of water.

Oxidizing Agents. The halogens such as fluorine, chlorine, and bromine are well-known oxidizing agents in the laboratory. Other powerful oxidizing agents are perchloric acid, chromic acid, and others. In addition to their corrosive properties, perchloric and chromic acids also present fire and explosion hazards on contact with organic compounds and other oxidizable substances. The hazards associated with the use of perchloric acid are especially severe. Perchloric acid must be handled only after thorough familiarization with recommended safe handling procedures. Strong oxidizing agents are to be stored and used in glass or other inert containers. Corks and rubber stoppers are not to be used to contain perchloric acid. Heating operations using significant quantities of these reagents are to be conducted in fiberglass mantles or sand baths rather than oil baths.

Handling Procedures

Prior to beginning a laboratory operation, each worker must be required to read and understand the information on the Material Safety Data Sheets (MSDS) of all corrosive chemicals to be handled. The overall objective of worker protection is to minimize exposure of the operator to corrosive chemicals by taking all reasonable precautions. The following precautions must always be followed:

(1) Protect the hands and forearms by wearing appropriate chemical-resistant gloves, a laboratory coat and rubber apron to avoid any contact of corrosive chemical with any part of the body.

(2) Procedures involving corrosive chemicals that may result in the generation of corrosive gases, vapors, fumes, aerosols, and dusts, must be conducted in a hood or other suitable containment device.

(3) If a hood is used, it must have been evaluated previously to establish that it is providing adequate ventilation and has an average face velocity of not less than 75 linear ft per minute.

(4) To minimize hazards from accidental breakage of apparatus or spills in the hood, containers are to be stored in pans or trays made of unbreakable, chemical-resistant containers and apparatus must be mounted above trays of the same type of material.

(5) The laboratory worker must be prepared for possible accidents or spills. If a corrosive chemical contacts the skin, the area must be well irrigated with copious amounts of running water and followed by a safety shower.

(6) All eyewash equipment must provide a minimum of 15 min of continuous water flow and be accessible within 10 sec or 100 ft (30.5 m) from

the hazard. Flushing streams should rise to approximately equal heights and wash both eyes simultaneously with a minimum delivery of 0.4 gal/min. (ANSI Z358.1). The first few seconds after contact of a corrosive with the eye are most critical. Quick action and immediate flushing of the eyes may prevent permanent damage. Sometimes, the victim's eyelids may have to be forced open, so that the eyes can be flushed.

(7) All safety showers must have a height for the shower head of 82 to 96 in. (2.08-2.44 m) with the following spray pattern: a minimum diameter of 20 in. (50.8 cm) at 60 in. (1.52 m) above standing surface, and with a minimum delivery of 30 gal/min. (ANSI Z358.1).

(8) All spills must be cleaned up by personnel wearing suitable personal protective apparel and equipment.

(9) Contaminated clothing and shoes must be removed from the body. Salvaged clothes and shoes must be properly cleaned and decontaminated.

(10) Upon completion of the operation, the worker must clean the exterior of his gloves and rubber apron before removal. They will then be clean and ready for the next operation.

HANDLING PROCEDURES FOR TOXIC CHEMICALS

A toxic chemical is any chemical which causes adverse effects on living organisms. Toxicology is the study of such adverse effects. Toxicity is the capacity of a chemical to produce injury. In practical situations, the critical factor is the hazard or risk associated with chemical use. Hazard or risk assessment takes into account possible toxic effects from the use of a chemical in the quantity and in the manner proposed. In evaluating either a hazard or a risk, toxicity is but one factor to be considered. Two chemicals may possess the same degree of toxicity but present different degrees of hazard. For example, carbon monoxide is odorless, colorless, and nonirritating to the eye, nose, and throat. On the other hand, ammonia has a pungent odor and is an eye, nose, and throat irritant. Yet, by comparison, ammonia, with the warning properties, presents a lesser degree of hazard than carbon monoxide, which could cause death.

Dose and Duration of Exposure

All toxicological considerations are based on the dose-response relationship. The toxic potency of a chemical is ultimately defined by the relationship between the dosage of the chemical absorbed and the response produced in a biological system. In toxicity testing, the death of an experimental animal is the commonly chosen endpoint. If the only variable considered is the number of deaths, it is then possible to use the concept of the lethal dose (LD). The dose which produces death in 50% of the experimental animals is commonly abbreviated as LD_{50}, and the dose is expressed as amount per unit of body weight such as mg/kg. The LD_{50} is the concentration that kills half of the exposed animals. Yet, the other 50% are not necessarily in good health. The LD_{50} values should also be accompanied by an indication of the species of experimental animal used, the route of administration for the chemical, and the time period over which the animals were observed.

If the experiment has involved inhalation as the route of exposure, the dose to the animal will be expressed as the concentration of the chemical in the air such as mg/m3. In this case the term LC_{50} is used to designate the concentration in air expected to kill 50% of the animals exposed for the specified length of time.

Route of Exposure

In discussing toxicity, it is necessary to describe how a chemical gains entrance into the body and then into the bloodstream. Once absorbed into the bloodstream, a chemical may elicit general effects or possibly the toxic effects will localize in specific tissues or target organs. The major routes by which chemicals gain access to the body are through the lungs (inhalation), the skin (topical absorption), and the gastrointestinal tract (ingestion).

Inhalation is by far the most important route of entry for chemicals into a living organism. Any airborne chemical can be inhaled. The total amount of chemical absorbed via the respiratory tract depends upon its concentration in the air, the duration and frequency of exposure, and the rate of breathing. The solubility of the gases and vapors in water is generally the major characteristic determining the relative toxicities of chemicals.

The second important route of entry is absorption through either intact or abraded skin. The absorption of organic compounds may follow surface con- tamination of the skin or of clothes, or it may occur directly from the vapor phase. Temperature elevation may increase skin absorption by increasing vasodilation. If the skin is damaged by scratching or other abrasion, the normal protective barrier to absorption is lessened and enhanced penetration may occur.

The problem of ingesting chemicals is not widespread, but accidental swallowing does occur. Ingestion of inhaled chemicals can also occur because chemicals deposited in the respiratory tract can be carried out to the throat by the action of the ciliated lining of the respiratory tract. These chemicals are then swallowed and significant absorption from the gastrointestinal tract may occur.

Action of Toxic Chemicals

The toxic action of a chemical can be arbitrarily divided into acute and chronic effects. Acute exposure and acute effects involve short-term, high concentrations and immediate results of some kind resulting in illness, irritation, or death. Acute exposures, typically, are sudden, severe, and characterized by rapid absorption of the offending chemical. For example, inhaling high concentrations of carbon monoxide or swallowing a large quantity of sodium cyanide produces acute poisoning. The critical period occurs suddenly for death or survival of a victim. Such incidents generally involve a single exposure in which the chemical is rapidly absorbed and in turn damages one or more of the vital organs. The effect of a chemical hazard is considered acute when it appears with little time lag, such as within minutes or hours.

In contrast to acute effects, chronic effects are characterized by either symptoms of a disease or the development of an actual disease of long duration. Frequently, reoccurrences slowly develop. The term chronic exposure relates to continual exposure to substances throughout a working lifetime. Chronic effects can be produced by exposure to a harmful material that produces irreversible damage such that the injury accumulates, rather than the poison itself. The symptoms in chronic poisoning are usually different from those seen in acute poisoning caused by the same toxic agent. For example, acute benzene poisoning affects the central nervous system, or may result in death, but chronic benzene poisoning affects the blood cell production capability of the bone marrow.

Effects of Exposure to Toxic Chemicals

The spectrum of undesired effects of chemicals is broad. Some chemicals are deleterious and others are not. One distinction between types of effects is made on the general location of action. Local effects refer to those that occur at the site of first contact between the chemical and the biologic system. Examples of local effects are demonstrated by the corrosive actions of acids and bases, and this has been discussed elsewhere. Systemic effects require absorption and distribution of the chemical to a site distant from its entry point. Most chemicals produce systemic effects. For some chemical, however, both local and systemic effects can be observed. For example, tetraethyl lead produces local effects on the skin at the site of absorption. Tetraethyl lead is then transported through the blood system to produce its typical effects on the central nervous system and other systems.

Asphyxiation is a common effect caused by some gases and vapors. Asphyxiants exert their effects by interfering with the supply of oxygen. Simple asphyxiants are physiologically inert gases that act by diluting atmospheric oxygen below that required to maintain blood levels sufficient for normal tissue respiration. Some common examples are the inert gases, hydrogen, nitrogen, and low-molecular-weight alkanes. Chemical asphyxiants, on the other hand, through their direct chemical action, either prevent the uptake of oxygen by the blood, interfere with the transportation of oxygen from lungs to the tissues, or prevent normal oxygenation of tissues even though the blood is well oxygenated. Carbon monoxide prevents oxygen dissociation by preferentially combining with hemoglobin, while hydrogen cyanide inhibits enzyme systems to utilize molecular oxygen.

Irritation is another toxic effect caused by gases and vapors. Irritation involves some sort of aggravation of whatever tissue the chemical comes in contact with. Direct contact of some chemicals with the face and upper respiratory tract affects the eyes, the tissues lining the nose, and the throat. Ammonia and chlorine are classic examples of irritant gases. Bronchoconstriction or the feeling of an inability to breathe occurs immediately on inhalation. Both gases are well tolerated in that, unless the concentration is sufficient to cause death, the acute effects do not result in chronic poisoning.

A variety of chemicals produce damage to the cells of the airways and alveoli. The resulting increase in permeability may lead to the release of fluids into the lungs, producing an edema. Ozone and nitrogen dioxide are examples of toxic chemicals that produce cellular damage. The water solubility of these chemicals is sufficiently low that the main site of action is in the lower respiratory tract. Phosgene is another irritant capable of producing delayed pulmonary edema. The moisture of the respiratory tract hydrolyzes phosgene to hydrochloric acid and carbon dioxide. Typically, a delay of approximately 24 hr lapses between exposure to phosgene and the development of symptoms. Individuals exposed to phosgene should be under medical surveillance for at least 48 hr.

Inhalation of solid particles or chemical dusts may have some health effects. A certain amount of filtration by the upper respiratory tract can prevent the larger particles from ever getting into the lung. Particles smaller than 10 µm in diameter can be inhaled and readily reach the deep lung. Once inhaled and deposited into the lung, the particles can remain without causing any damage. Inert dusts such as calcium carbonate and barium sulfate are considered relatively harmless unless the exposure is severe and long-term. Inert dusts are sometimes called nuisance dusts. They may cause radiopaque deposits in the lung that are visible on X-ray film, but they produce little or no tissue reactions unless the exposure is overwhelming.

Pneumoconiosis is the term applied to a class of diseases caused by the accumulation of dusts in the lungs. Pneumoconiosis associated with inert dusts are potentially reversible and sometimes called benign pneumoconiosis. Far more serious is the effect of insoluble particles which cause fibrotic changes in the lung. Fibrotic changes are produced by materials such as free silica which produce the typical silicotic nodule or small area of scar-like tissue. In addition, asbestos fibers produce typical fibrotic damage to lung tissue as well as cancerous lung changes.

Chronic pulmonary disease can result from inhalation of a variety of materials which appear to act wholly or partly through an allergic response. The underlying mechanism is demonstrated by the presence of antibodies to specific components of the inhaled materials. In some instances, these reactions are caused by spores of molds or by bacterial contaminants. In other instances, as in the case of cotton dusts, they appear to be related to components of the material itself. An example of an organic dust that can produce allergic-like symptoms on inhalation is toluene diisocyanate. Toluene diisocyanate is widely used in the manufacture of polyurethane plastics.

It is a well-established fact that exposure to some chemicals can produce cancer in laboratory animals and humans. One definition of a carcinogen is a substance with the capability to induce a malignant tumor in an animal following a reasonable exposure. Benzene exposure has been associated with blood dyscrasias. Exposure to benzene may progress to leukemia. Coal tar and various petroleum products have been identified as skin and subcutaneous carcinogens. Vinyl chloride monomer has been identified as a cause among some workers of cancer of the lining of the liver. Inorganic salts of metals, such as beryllium and chromium, are associated with cancer of the respiratory tract.

Within the scope of this space, it is impossible to discuss all the specific toxic effects of various classes of chemicals. Furthermore, new chemicals are discovered every day, and along with them, new toxic effects. Today's researchers hold a limited knowledge of mutagenic and embryotoxic chemicals. Extensive research remains to be done on the classification and categorizing of the effects of millions of chemicals. Prudence and care are the most important factors in the safe handling of chemicals regardless of whether or not specific toxicity is known. Despite the potential hazards of thousands of chemicals each year, most injuries from chemicals are due to those in daily and general use.

Handling Procedures

Prior to beginning a laboratory operation involving any toxic chemical, each worker is strongly advised to consult a toxicology reference that lists toxic properties of chemicals, and to become knowledgeable of the proper handling procedures. Each laboratory worker's plans for experimental work must be approved by the laboratory supervisor. Consultation with the chemical hygiene officer is appropriate to ensure that the toxic chemical is effectively contained during the experiment.

It would serve the interests of the laboratory worker and the organization to maintain an accurate record of the amounts of the toxic chemicals used, dates of use, and names of users. The chemical hygiene officer is responsible for ensuring that accurate records are kept. The following precautions must always be followed:

(1) Any chemicals having high chronic toxicity must be stored in a ventilated storage area in a secondary tray or container having sufficient capacity to contain the chemical should the primary container accidentally break.

(2) Storage areas for substances in this category must have limited access, and special signs must be posted. Storage areas must be maintained under negative pressure with respect to surrounding areas.

(3) All containers of highly toxic chemicals must have labels that identify the contents and include a warning label to describe the type of toxicity if known, e.g., CANCER CAUSING AGENT.

(4) All experiments with and transfers of highly toxic chemicals or mixtures must be performed in an area designated for the use of such chemicals or mixtures. This designated area can be a portion of the laboratory, an exhaust hood, or a glove box. Designated areas must be clearly marked with a conspicuous sign, e.g., WARNING: TOXIC CHEMICAL IN USE.

(5) The type of glove box used to handle highly toxic chemicals must be of negative pressure with a ventilation rate of at least two volume changes per hour and the pressure of at least 0.5 in of water lower than that of the external environment. The exit gases from the glove box must pass through a trap or high efficiency particulate air (HEPA) filter.

(6) Laboratory vacuum pumps used with highly toxic chemicals must be protected by high-efficiency scrubbers or HEPA filters and vented into an exhaust hood. Motor-driven vacuum pumps are recommended because they are easy to decontaminate. Obviously, decontamination of a vacuum pump must be carried out in an exhaust hood.

(7) Laboratory workers handling highly toxic chemicals must wear proper gloves to avoid any skin contact. In some cases, the laboratory worker or the chemical hygiene officer may deem it advisable to use other personal protective apparel and equipment.

(8) Surfaces or bench tops on which highly toxic chemicals are handled must be protected from contamination by using chemically resistant trays or pans that can be decontaminated after the operation. Alternatively, disposable absorbent paper liners may be used.

(9) Prior to leaving the designated area, the laboratory worker must take precautions to clean the exterior of gloves and apron, remove any used protective apparel, and thoroughly wash hands, forearms, face, and neck.

(10) If disposable apparel or absorbent paper liners have been used, these items must be placed in the proper waste container. Nondisposable protective apparel must be thoroughly washed, and containers of disposable apparel and paper liners must be incinerated.

(11) By the end of the work day and before leaving the laboratory, all surfaces and bench tops on which highly toxic chemicals have been handled must be properly decontaminated. If chemical decontamination is to be used, a method should be chosen that can reasonably be expected to convert essentially all of the toxic materials into nontoxic materials.

(12) Collect wastes and other contaminated materials from all experiments involving highly toxic chemicals together with the washings from flasks and other containers used. All laboratory wastes must be placed in closed, suitably labeled containers for proper disposal.

9   STORAGE OF CHEMICALS

A wide range of possibilities exists for the storage of chemicals. For the safe and proper storage of laboratory chemicals, two principles apply: (1) Maintain an inventory control of chemicals, and (2) Store chemicals by class and by compatibility, i.e., mutually incompatible chemicals must be segregated from each other.

INVENTORY CONTROL

An inventory of all hazardous, as well as nonhazardous, chemicals must be conducted and reported to the chemical hygiene officer. From the inventory information, a list of the potential hazards and appropriate precautionary measures can be developed. This list will serve as a helpful guide to the preparation of any chemical hygiene plan. Proper inventory control consists of the following elements:

(1) Assign one person the responsibility of inventory control. In many organizations, the principal investigator or project director may initiate an order and have the chemicals delivered directly to the individual initiating the order. Lack of complete inventory information may result not only in a deficient chemical hygiene plan but in a deficient chemical emergency response plan as well.

(2) Provide sufficient storage space for chemicals. Often the provision of adequate storage space is given little consideration in the design of laboratory buildings. Lack of sufficient storage space can create hazards due to overcrowding, storing together of incompatible chemicals, and poor housekeeping. Adequate, properly designed and ventilated storage space must be available to ensure safety of personnel and the protection of property.

(3) Limit the amount of reserve supply. Centralize storage of chemicals with only: a month's supply on site; a week's supply in the building; and a day's supply in the laboratory.

(4) Ordering in large quantities reduces the unit cost of many chemicals. Obviously, large shipments of a chemical require a large storage space. Yet, keeping a hazardous chemical, such as diethyl ether, beyond its expiration date is potentially disastrous. The lesser unit cost both of large quantity and of annual bid purchases may be outweighed by the cost of rcbuilding a laboratory.

(5) Today, many chemical suppliers assign an expiration date to each chemical. The expiration date must be strictly observed. The expired chemical must either be marked for proper disposal or destruction.

(6) Generally, a first-in, first-out system of stock keeping should be used. However, the expiration date on the containers of the inventoried chemical must also be considered.

(7) If a chemical does not carry an expiration date, assign one to each chemical and date the container clearly at the time of receipt or preparation. The expiration date should not be longer than 1 year. Again, an expired chemical is either marked for proper disposal or destruction, or, if warranted, given a new expiration date.

(8) Stored chemicals must be examined at periodic intervals depending on the reactivity or instability of the chemicals. At this time, chemicals kept beyond their expiration date, or those which have deteriorated, questionable labels, leaks, corroded caps, or any other problem, must be marked for proper disposal or destruction.

STORAGE ACCORDING TO CHEMICAL COMPATIBILITY

In chemical history, numerous accidents and disasters were the direct results of improper chemical storage. Improper chemical storage implies the use of alphabetical listing of chemicals for convenience or a numbering system that disregards the compatibility of the chemicals. Examples in Table 9-1 illustrate how pairs of chemicals may be stored adjacent to each other alphabetically but are in fact chemically incompatible. In case of an accidental spill, fire, or other mishap in a storage area, the potential of mixing reactively incompatible pairs of chemicals may have disastrous results.

The principle of storage of chemicals according to compatibility consists of the following elements:

(1) Having completed the inventory, all chemicals on site must be segregated into broad classes such as acids, bases, flammables, oxidizers, etc. After completion of this task, a good chemist will immediately realize that not all chemicals within any single broad

Table 9-1   Dangers of Storing Chemicals in Alphabetical Order

Chemicals Stored Together	Possible Reactions
Acetic acid + acetaldehyde	Acetic acid will cause the polymerization of acetaldehyde, thus releasing large amounts of heat
Acetic anhydride + acetaldehyde	Reaction can be violently explosive
Aluminum metal + ammonium nitrate	A potential explosive
Barium + carbon tetrachloride	A potential violent reaction
Cadmium chlorate + cupric sulfide	Will explode on contact
Carbon + bromate, chlorate, or Iodate	A potential explosive combination detonated by heat, percussion, or friction
Copper + bromate, chlorate, or Iodate	A potential explosive combination detonated by heat, percussion, or friction
Ferrous sulfide + hydrogen peroxide	A vigorous, highly exothermic reaction
Lithium metal + maleic anhydride	Maleic anhydride decomposes explosively
Mercury(II) oxide + phosphorus	Percussion may ignite this mixture
Methyl alcohol + lead perchlorate	An explosive mixture if agitated
Methyl alcohol + mercury(II) nitrate	May form mercury fulminate, an explosive
Nitric acid + phosphorus	Phosphorus will burn spontaneously
Potassium metal + potassium peroxide	The peroxide can oxidize the metal to incandescence
Silver metal + sulfur	An explosive mixture
Silver oxide + tartaric acid	An explosive mixture
Sodium nitrate + sodium thiosulfate	A dry mixture can result in explosion
Stannic chloride + turpentine	A flame-producing, exothermic reaction

group are mutually and chemically compatible. For example, some acids are oxidizing agents, e.g., nitric acid; others are reducing agents, e.g., glacial acetic acid. Obviously, two incompatible acids must be segregated in storage. Through an examination of a class of chemicals, one discovers many chemicals within a single broad class are not mutually compatible.

(2)   Within a broad class of chemicals, further segregation into chemically related and compatible groups is necessary. The two suggested lists below represent some of the chemically related and compatible groups. Readers are reminded that these two lists are by no means

complete. These lists are intended to cover only the more commonly used laboratory chemicals.

Examples of Inorganic Compatible Groups
- Acids (except HNO_3).
- Amides, azides, nitrates (except NH_4NO_3), nitrites, HNO_3.
- Arsenates, cyanides.
- Borates, chromates, manganates, permanganates.
- Carbides, nitrides, phosphides, selenides, sulfides.
- Carbonates, silicates.
- Chlorates, chlorites, hypochlorites, perchlorates, perchloric acid.
- Halides, halogens.
- Hydroxides, oxides.
- Phosphates, sulfates, sulfites, thiosulfates.
- Metals, hydrides.
- Phosphorus, phosphorus pentoxide, sulfur.

Examples of Organic Compatible Groups
- Acids, anhydrides, peracids.
- Alcohols, glycols.
- Aldehydes, esters.
- Amines, amides, imines, imides.
- Azides, hydroperoxides, peroxides.
- Cresols, phenols.
- Ethers, ketenes, ketones.
- Epoxy compounds, isocyanates.
- Hydrocarbons, halogenated hydrocarbons.
- Nitriles, sulfides, sulfoxides.

(3) Assign all chemicals into chemically related and compatible groups and segregate each group by physical barriers or by distance. The group arrangements of chemicals depends on the size of the storage area, the quantities handled, and the nature of the problem.

(4) Access to any chemical storage area must be strictly limited to the selected personnel with the direct responsibility for inventory control and segregation of chemicals into compatible groups.

PROCEDURES FOR THE STOREROOM/STOCKROOM

The following chemical handling procedures are recommended for storerooms and stockrooms:

(1) Storerooms and stockrooms are defined as chemical storage areas. The storage areas must be well ventilated, with both floor and ceiling vents. If storage of opened containers is permitted, additional local exhaust ventilation and the use of outside storage containers or spill trays is mandatory.

(2) All chemicals should be stored only to eye level. Chemicals are never to be stored on floors. Whenever possible, solids must be stored above liquids in the same tier of shelves. A retaining shock cord or some similar device across the open face of the shelf may be used to prevent accidental toppling of containers of chemicals.

(3) Stockrooms are similar to central storerooms except that the quantities of chemicals are usually much smaller. Stockrooms are not to be used as preparation areas because of the possibility that an accident may occur and thereby unnecessarily contaminating a large quantity of virgin chemicals. Preparation or repackaging of chemicals is to be performed in an area other than the stockroom.

(4) Stockrooms should be conveniently located and open during normal working hours to ensure laboratory workers are not tempted to store excessive quantities of chemicals in their individual work area. If the employment of a full-time stockroom clerk is economically unfeasible, one readily available employee should be assigned the responsibility of stockroom safety and inventory control.

(5) Supplies for cleanup spills, appropriate for chemicals stored, must be available in close proximity to those chemicals. The same principle applies to other safety equipment such as fire extinguishers.

(6) Each storeroom or stockroom must have at least two exits. Each exit is to be protected with automatic sprinkler system of water or other suitable fire extinguishing agent.

(7) Each storeroom or stockroom must have at least one escape self-contained breathing apparatus. All who work in or even occasionally enter the area must be trained in the use of this equipment.

(8) Light switches, electric outlets, motors, and other electric devices must be both nonsparking and explosion proof.

(9) The floor of the storeroom or stockroom must be diked to confine any potential liquid chemical spills. The diked area should be fitted with a special drain leading to a separate pond for treatment prior to waste disposal.

(10) In addition to the standard safety equipment such as eyewash fountain and safety shower, the room must be equipped with fire, smoke, flammable vapor, and temperature alarms. Personnel working in the storeroom or stockroom as well as other employees in the adjacent area must be trained to distinguish between and among the alarm signals.

STORAGE OF FLAMMABLE CHEMICALS

The best method of controlling potential fire hazards of flammable liquids is to centralize storage of bulk quantities. The following procedures are recommended for flammable chemical storage:

(1) The most effective way to minimize the impact of a hazard is isolation. Therefore, a storage and dispensing facility for flammable liquids is best located in a special building separate from the main building.

(2) If the storage and dispensing room must be located in the main building, the preferred location is a cutoff area on the at-grade level with at least one exterior wall. Cutoff is a fire protection term meaning "separated from other areas by fire-rated construction."

(3) For obvious reasons, a storage facility for flammable liquids should never be placed on the roof of a building, located on a below-grade level, an upper floor, or in the center of the building. The locations listed are less accessible for fire-fighting and potentially dangerous to the safety of the personnel in the building.

(4) The walls, ceilings, and floors of an inside storage room for flammable liquids must be constructed of materials of at least a 2-hr fire resistance, and with self-closing fire doors (29 CFR 1910.106).

(5) All storage rooms for flammable liquids must have adequate mechanical ventilation controlled by a switch outside the door and explosion-proof lighting and switches. Other potential sources of ignition, such as burning tobacco and lighted matches, must be forbidden.

(6) Fifty-five gallon drums are commonly used to ship flammable liquids but are not intended as long-term indoor storage containers. The bung must be removed and replaced by an approved pressure and vacuum relief vent to protect against internal pressure buildup in the event of fire.

(7) If possible, drums should be stored on metal racks placed with the end bung openings toward an aisle and the side bung openings on top.

(8) The drums, as well as the metal racks, must be properly grounded. It is also necessary to provide bonding to metal receiving containers to prevent accumulation of static electricity.

(9) Drip pans with flame arresters must be installed or placed under faucets. Drum faucets of plastic construction are not generally acceptable due to chemical action on the plastic materials.

(10) A dispensing device safer than the faucet is the hand-operated rotary transfer pump. Such pumps have metering options and permit immediate cutoff control to prevent overflow and spillage. It can be

reversed to siphon off excess liquid in case of overfilling and be equipped with drip returns to return excess flammable liquid to the drum.

(11) Whenever feasible, quantities of flammable liquids greater than 1 L should be stored in metal containers. Portable approved safety cans are one of the safest methods of storing flammable liquids. These cans are available in a variety of sizes and materials. Equipped with spring-loaded spout covers, they can open to relieve internal pressure when subjected to a fire. Leakage is prevented if the can tips over. Some are equipped with a flame arrester.

(12) Small quantities of flammable liquids can be stored in ventilated storage cabinets made of 18-gauge steel and having riveted and spot-welded seams. Such cabinets are of double-wall construction and have a 1.5-in. air space between the inner and the outer walls. The door is 2 in. above the bottom of the cabinet and the cabinet is liquid tight to this point. It is provided with vapor-venting provisions. Some models are equipped with doors that close automatically in the event of fire.

(13) Other considerations in the storage of flammable liquids in the laboratory include ensuring that the aisles and exits are not blocked in the event of fire; that accidental contact with any strong oxidizing agents is not possible; and that sources of ignition are excluded.

STORAGE OF OXIDIZERS

Mineral acids, including those recognized as strong oxidizers — such as nitric acid, perchloric acid, and sulfuric acid — must be segregated from flammable chemicals. Such mineral acids must be stored in separate rooms, separate cabinets, and in break-resistant containers if large glass bottles must be stored in proximity of flammable chemicals.

STORAGE OF WATER-REACTIVE CHEMICALS

Water-reactive chemicals react with water to evolve flammable or explosive gases. For example, potassium and sodium metals, and many metal hydrides, react on contact with water to produce hydrogen. These reactions evolve sufficient heat to ignite the hydrogen with explosive violence. Certain polymerization catalysts, such as aluminum alkyls, react and burn violently on contact with water.

Storage facilities for water-reactive chemicals must be constructed to prevent their accidental contact with water. This is best accomplished by eliminating all sources of water in the storage area, e.g., areas where large quantities of water-reactive chemicals are stored should never have automatic water sprinkler systems. Furthermore, such facilities must be of fire-resistant construction. Flammable chemicals should never be stored in the same area.

STORAGE OF CORROSIVE CHEMICALS

The major classes of corrosive chemicals are mineral acids, bases, dehydrating agents, and oxidizing agents. As discussed under storage of oxidizers, all solid and liquid oxidizers should be segregated from flammable chemicals, stored in separate rooms, separate cabinets, and in break-resistant containers if large quantities must be stored in proximity to flammable chemicals. The following additional recommendations must be followed:

(1) Prevent corrosion of metal shelves or oxidization of wooden shelves by providing acid-resistant trays under containers of corrosive chemicals.
(2) Emergency water and safety equipment, such as eye and face wash fountains and safety showers, must be available in all of these storage areas since many of these chemicals are corrosive to human tissues.
(3) Acids are incompatible with bases, and oxidizing agents with reducing agents. These groups of chemicals must be segregated from each other. If they have to be in close proximity to each other, they should be placed in separate cabinets and in break-resistant containers.
(4) Supplies for the cleanup of spills and neutralizing agents appropriate for the corrosive chemicals being stored must be available near these chemicals.

STORAGE OF TOXIC CHEMICALS

The following procedures are recommended for the storage of toxic chemicals:

(1) Toxic chemicals must be segregated from other chemicals and stored in a well-defined or identified area that is cool, well-ventilated, and away from light, heat, acids, oxidizing agents, and moisture.
(2) Chemicals known to be highly toxic, including those classified as carcinogens, embryotoxins, etc. must be stored in ventilated storage areas in unbreakable and chemical-resistant secondary containers. Adequate ventilation is of particular concern for highly toxic chemicals that have a high vapor pressure.

(3) The storage of unopened containers of toxic chemicals normally presents no problems. However, opened containers must be closed with tape or other sealant before being returned to the storeroom and they must never be returned unless some type of local exhaust ventilation is available.

(4) Only minimum working quantities of highly toxic chemicals can be allowed in the work area. Containers of such chemicals must carry an appropriate warning label, e.g., CAUTION: HIGHLY TOXIC CHEMICAL.

(5) Storage areas for highly toxic chemicals must exhibit a sign warning of the hazard and have limited access.

(6) An inventory of highly toxic chemicals must be maintained. This inventory may be required by federal and state regulations, e.g., regulated carcinogens.

10   LABORATORY VENTILATION

Ventilation is a key method for reducing employee exposures to airborne contaminants resulting from laboratory operations. Ventilation can be used either to dilute contaminants to safe levels or to capture and remove contaminants at their sources before pollution of the working environment occurs. Occupational Safety and Health Act standards set legal limits on employee exposures and focus attention on the need to identify and reduce hazardous exposures.

In most laboratories, exhaust ventilation is the responsibility of the safety department and plant engineering staff. Often the safety department identifies the potentially harmful exposures and the plant engineer designs and installs the system hardware. Whether ventilation is used where needed and whether it properly provides adequate protection depends on the knowledge and skill of the safety department and plant engineering staff.

DILUTION VS. LOCAL EXHAUST

Dilution occurs when contaminants mix with air flowing through the work area. Either natural or mechanically induced air movement dilutes contaminants. The major disadvantages of dilution ventilation are that large volumes of dilution air may be needed and employee exposures are difficult to control near the undiluted contaminant. Dilution ventilation is also called general ventilation.

Local exhaust systems capture or contain contaminants at their source before escaping into the workroom environment. A typical system consists of ducts, one or more hoods, an air cleaner if needed, and a fan. The big advantage of local exhaust systems is that they remove contaminants rather than just dilute them. A second major advantage is that these systems require less airflow than dilution ventilation systems in the same application. The total airflow is important for the heating or cooling of plants. Heating and air conditioning costs are an important operating expense.

Although both types of ventilation are useful, local exhaust is usually preferred, if it is feasible. Advantages include more positive control of employee exposures and lower overall airflow requirements. On the other hand, dilution ventilation systems are usually less expensive to install and operate if heating costs during the winter are not excessive.

OSHA VENTILATION STANDARDS

Ventilation is required in OSHA standards for two reasons: to control employee exposures to potentially harmful materials and to prevent fire or explosion hazards. Within both categories there are two types of requirements:

Performance Standards set a goal, such as maintaining exposure to airborne contaminants below the OSHA permissible exposure limits, and allow the employer to meet that goal in the best way, e.g., the OSHA Laboratory Standard being discussed in this book. Readers are reminded that the OSHA Laboratory Standard is a performance standard.

Specification Standards require the installation of a ventilation system for certain processes and a compliance with specific design requirements. For example, whenever zinc-bearing or zinc-coated metal is welded indoors local exhaust ventilation must be provided. The ventilation system must move enough air to maintain an air velocity of 100 ft/min through the welding zone when working the maximum distance from the hood opening.

Specification standards do not give the employer much latitude, but they generally identify situations that are recognized as hazardous enough to warrant controls. Some standards are combinations of performance and specification criteria. For example, when welding or torch cutting on cadmium-bearing or cadmium-coated metal is done indoors, either local ventilation or air-line respirators are required unless atmospheric tests under the most adverse conditions have established that the worker's exposure is within the acceptable concentrations allowed by OSHA.

LOCAL EXHAUST SYSTEMS

Dilution exhaust systems lower the contaminant concentrations by dilution with fresh air but do not reduce or eliminate the total amount of contaminant released. In contrast, local exhaust systems capture air contaminants at or near their source of generation before they are dispersed into the workroom air. Therefore, local exhaust is frequently the preferred method for controlling airborne concentrations of potentially hazardous materials. A typical local exhaust system consists of four components: hoods, ducts, an air cleaner, and a fan.

Hoods are the openings into the ventilation system where contaminants are either captured or retained by flowing air currents. The hoods are the most important part of the system. Different hoods work in different ways: some reach out and capture contaminants; others catch contaminants thrown into the hood; still others contain contaminants released inside the hood and prevent them from escaping into the work area. *Ducts* are a network of piping that connects the hoods and other systems components. An *air cleaner* is a device that removes airborne materials carried in the exhaust air. Air cleaners designed to remove both solid particles and gaseous contaminants are available. A *fan* is the air-moving device that provides the energy to draw air and contaminants into the exhaust system by inducing a negative pressure or suction in the ducts leading to the hoods. The fan converts electrical power into negative pressure and increased air velocity.

A ventilation system is usually designed to fit the existing operations. A hood shape and location are chosen depending on the source of contamination. The airflow volume into each hood is then determined from reference sources or from experiments with models of the hood. With this information the ducts and air cleaner can be sized. The fan size needed to draw the required amount of air, while overcoming friction and other resistance, can be determined. Now the system is ready for installation. Testing is done to determine whether it meets design criteria.

Pressure is the key concept in understanding how local exhaust systems work. Air starts moving because there is a difference in pressure between two points. Suction is the reduced pressure caused by the fan that draws the desired amount of air into the hoods and through the ducts and air cleaner. The amount of suction *(static pressure)* that the fan must generate for proper system performance represents the energy added to the system to overcome pressure losses from acceleration, hood entry, duct friction, and elbow turbulence as well as the resistance loss of air cleaner. Static pressure represents the potential energy of the system. The magnitude of the loss is proportional to the square of the air velocity that, when converted to pressure and expressed in units of inches of water, is called *velocity pressure*. Velocity pressure represents the kinetic energy in the system. *Total pressure* is the sum of static and velocity pressures and represents the total energy in the system.

TYPES OF HOODS

The hood is the most important part of a ventilation system. No local exhaust system will work properly unless enough of the contaminants are captured by the hoods so that the concentration of contaminants in the workroom air is below

Fan

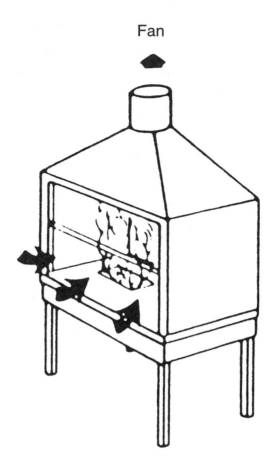

Figure 10-1   Enclosure type of hood.

permissible exposure limits. Both the design and location of the hoods are crucial in determining whether a system will work. A poor hood design may prevent the ventilation system from ever working adequately. It may also result in excessive power costs as fan size and speed are increased to compensate for the initial poor hood selection. There are three major types of hoods, each working on a different principle:

Enclosure Hood

This type of hood surrounds the contaminant source as much as possible. Contaminants are kept inside the enclosure by air flowing in through openings in the enclosure (Figure 10-1). The more complete the enclosure, the less airflow is needed for control. The design should distribute air within the enclosure to

Fan

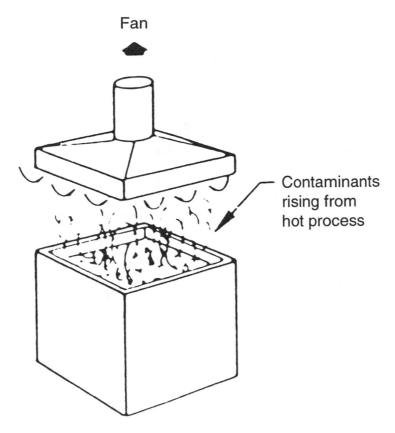

Contaminants
rising from
hot process

Figure 10-2   Receiving type of hood.

prevent accumulation of explosive or flammable vapor concentrations. Due to low exhaust rates, enclosures are usually the most economical hoods to install. Inward face velocities of 100-150 ft/min are typical.

Receiving Hood

Some processes generate a stream of contaminants in a specific direction, e.g., a furnace emitting a hot stream of air and gases rising above the unit. The ideal hood for this type of process is one that is positioned to catch the contaminants when they are being generated (Figure 10-2). A major limitation to the use of receiving hoods is that gases, vapors, and the very small particles which can be inhaled do not travel very far in air unless carried by moving air. Therefore, receiving hoods are not very useful for health protection ventilation systems unless the process emits quantities of hot air or air with sufficient velocity to carry the respirable contaminants into the hood.

Fan

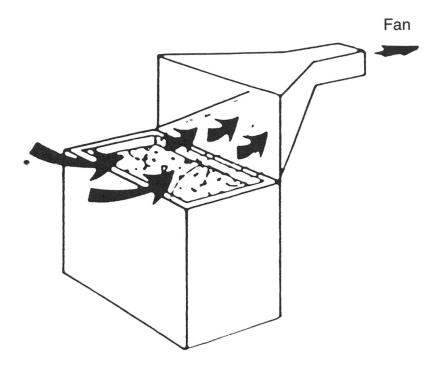

Figure 10-3   Capturing type of hood.

Capturing Hood

This type of hood reaches out to capture contaminants in the workroom air (Figure 10-3). Capture velocities range from 50-2000 ft/min depending on the process, the toxicity of materials released, and the conditions in the work area. They are widely used since capturing hoods can be placed alongside the contaminant source rather than surrounding it with an enclosure hood. The primary disadvantage is that large air volumes may be needed to generate an adequate capture velocity at the contaminant source. A second disadvantage is that the reach of most capturing hoods is limited to about 2 ft from the hood opening.

Selecting the proper hood depends on the contaminant. Gases, vapors, and minute particulates in the respirable size range have no independent motion; they move with the air around them. Contaminants are thus controlled by the regulations of the air with which the contaminants move. Particulates with aerodynamic diameter of 50 μm are too large to remain airborne and so rarely present an inhalation hazard. Hence controlling large particles is not usually important from a health standpoint. The important factor in hood selection is that each hood type works on a different principle. By confusing or violating these principles, the hood will not function properly in controlling contaminants.

TESTING OF HOODS

Every ventilation system for contaminant control must be checked periodically for proper operation. A newly installed system must be thoroughly tested to ensure it meets design specifications and to ensure airflow through each hood is correct, i.e., that the system is properly balanced.

From the viewpoint of the safety and health department, the periodic checks of system performance are important in protecting workers' health. After a system is installed and operating properly, the system's efficiency in containing or capturing contaminants may decrease for many reasons: loose fan belt; unauthorized damper adjustment; dust settled in ducts; and clogged air cleaners. Occasionally an existing system will become unbalanced when new ducts are added.

The important factor is not just how well the hardware is performing but how much protection the system provides. Although a system's airflow through hoods, air cleaner efficiency, fan speed, and power consumption may check out according to design specifications, the system is not performing adequately if air samples show excessive employee exposures. Perhaps the wrong type or size of hood was specified in the original design or there are uncontrolled sources of contamination in the workplace. Only by correlating ventilation system test results with air contaminant concentrations can one be sure that the ventilation system is providing sufficient protection.

Different tests are available to help determine how the system is operating. These tests are: smoke tube tracers; velocity at the hood opening; hood static pressure; Pitot tube duct velocity measurements, system static pressure tests; fan pressure, rotating speed, and power consumption; and periodic maintenance.

Smoke Tube Tracer Test

Smoke tubes for testing ventilation systems contain titanium tetrachloride or other chemicals which produce fumes by reaction with air blown through the tube. For use, the tube tips are broken off and air is blown through the tube with a rubber squeeze bulb. The smoke follows the air currents and its velocity shows how air is flowing into the hood or in the laboratory. Smoke tube tracer tests are useful in the following applications:

(1) *Contaminant Dispersion Pattern.* The correct hood location and size depend on where the contaminants originate and the velocity of the contaminants. Smoke tube tracers help identify the point where the initial velocity of contaminants away from the hood is dissipated so the airflow to develop the necessary hood capture or containment velocity at that point can be determined.

(2) *Capture Distance.* Smoke tube tracers can help to determine the approximate capture distance from a hood opening that the contaminant will be captured. Although smoke tubes are not as accurate as velometer readings, the tubes will help estimate the air velocity according to smoke behavior.

(3) *Contaminant Escape Pattern.* Since these phenomena occur randomly and in localized areas of enclosure openings, velometer readings may not identify them. Smoke tube tracers are the techniques of choice in such studies.

(4) *Hood Performance Demonstration.* Smoke tube tracers show workers how ventilation hoods function and what effects distance, damper settings, crossdrafts, and other factors have on hood performance.

Hood Velocity Measurements

Although smoke tube tracer tests are convenient and easy to perform, the results are qualitative. If the smoke flows into the hood with reasonable velocity, the hood is judged to be adequate. In order to obtain numerical air velocity readings, the following tests are available:

(1) *Deflecting vane velometer* is the most common field velocity meter. It is available for measuring velocity from zero to 10,000 ft/min in a variety of hood openings. Typical applications include measuring capture velocity outside of capturing hoods; face velocity for enclosures; and slot velocities.

(2) *Heated wire anemometer* works on the principle that the resistance of a heated wire changes with temperature variations. Air moving over the heated wire changes its temperature depending on the air velocity. Applications are similar to the deflecting vane velometer.

(3) *Rotating vane anemometer* is useful in large openings such as doorways or large ventilated booths. A rotating vane anemometer is not recommended for small areas where the instrument fills an appreciable portion of the opening.

Hood Static Pressure Test

Hood static pressure is the amount of suction in the duct near the hood. A certain hood static pressure is needed to draw the correct amount of air into the hood. If a malfunction such as a plugged duct or loose fan belt happens to the ventilation system, the suction available at the hood declines and the hood draws less air. Any change in hood static pressure is easily measured using a manometer.

DUCT VELOCITY MEASUREMENT

Although hood velocity or hood static pressure measurements are adequate for periodic tests, neither gives accurate determination of airflow through the system. The most accurate way to measure airflow in ducts is to perform a "Pitot traverse" of the duct. A Pitot tube is used to measure the air velocity at a number of points across the duct cross-section. Major applications of the Pitot traverse are for initial testing and balancing of new systems as well as for troubleshooting systems performing inadequately.

A Pitot traverse involves measuring the velocity at a number of points across the duct area since velocity distribution is not uniform within the duct. As a rule of thumb the average duct velocity is about 90% of the centerline velocity. A traverse is necessary whenever turbulent or stratified airflow in the duct is suspected. The number and location of measuring points within the duct depend on the size and shape of the duct.

LABORATORY FUME HOODS

A laboratory fume hood is a ventilated enclosure where harmful chemicals can be handled safely. The purpose of a laboratory fume hood is to capture and contain toxic contaminants from escaping into the laboratory. This is accomplished by drawing contaminants within the hood's work area away from the operator, so that inhalation and contact are minimized. Air flow in the hood is achieved by an exhaust blower which pulls air from the laboratory into its duct work system. The pull at the work-surface opening, also called a face, of the hood creates an air flow that can be quantified in terms of a capture velocity sufficient to contain toxic chemicals. Contaminated air is then exhausted through its duct system.

There are two major kinds of laboratory fume hoods: general purpose and special purpose. As the title implies, general purpose laboratory fume hoods are suitable for general work with chemicals that do not require special handling.

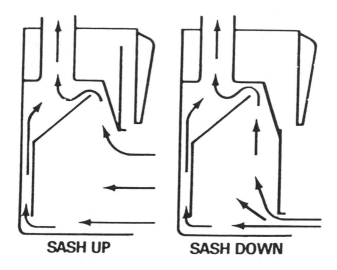

Figure 10-4   Conventional fume hood.

Special purpose laboratory fume hoods are suitable for procedures that create reactivity, health hazards, or other dangers.

General Purpose Fume Hoods

General purpose laboratory fume hoods are divided into three categories: conventional, by-pass air, and add-air, and each of these is described below:

Conventional

This is a basic enclosure with a movable front sash and an interior baffle (Figure 10-4). The sash may be fully opened to accommodate a variety of apparatus inside the hood, and the baffle guides exhaust air to follow specific flow patterns within the hood. The conventional hood is generally the least expensive, but its performance depends largely on sash position.

Bypass-Air

The bypass-air hood is quite similar to a conventional hood, except that it is designed to permit some exhaust air to bypass the sash closure (Figure 10-5).

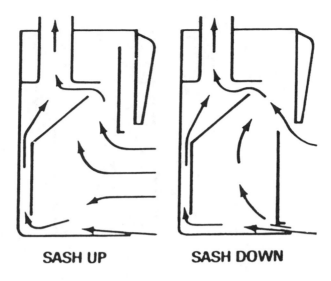

SASH UP   SASH DOWN

Figure 10-5   Bypass-air fume hood.

This eliminates the basic drawback of the conventional hood. The bypass-air hood is more desirable for use with fragile apparatus and instrumentation since the face velocity does not reach levels which might be detrimental to experimentation.

Add-Air

The add-air hood provides a means of introducing outside air to the hood exhaust and limits the percentage of tempered air removed from the laboratory (Figure 10-6). This type of hood is also known as: induced air, auxiliary air, balanced air, makeup air, and controlled air hood. Auxiliary air, when properly applied, can in some instances provide energy savings.

Special Purpose Fume Hoods

Procedures that create reactivity, health hazards, or other dangers, require the use of special purpose fume hoods. While there are almost as many specialty purpose hoods on the market as there are special purposes, it is beyond the scope of this chapter to discuss them all. Special purpose fume hoods are of the bypass-air type and the most common are the perchloric acid fume hood and the radioisotope fume hood.

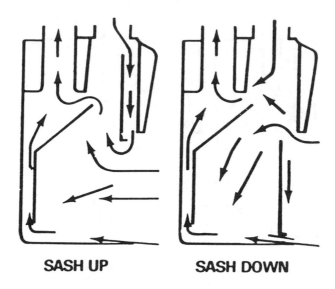

SASH UP   SASH DOWN

Figure 10-6   Add-air fume hood.

Perchloric Acid Fume Hood

Due to the potential explosion hazard of perchloric acid when combined with organic materials, this hood type must be used for perchloric acid digestion. Perchloric acid fume hoods have integral bottoms, coved interiors, and a drain. Wash-down features are required since the hood and duct system must be thoroughly rinsed after each use to prevent the accumulation of reactive residue.

Radioisotope Fume Hood

Hoods used for radioactive applications should have integral bottoms and coved interiors to facilitate decontamination. Radioisotope fume hoods should be strong enough to support lead shielding bricks, in case they are required. They should also be constructed to facilitate the use of high-efficiency particulate filter (HEPA) filters.

Face Velocity of Fume Hoods

Regulatory compliance agencies and other trade groups have established guidelines relating to the toxicity of various chemicals. These toxicity levels are identified as Threshold Limit Values (TLVs). In general, the lower the TLV number, the more toxic the chemical, and the higher the face velocity will be

**Table 10-1   Recommended Face Velocity of
Laboratory Fume Hoods**

Classification	Face Velocity
General Purpose Laboratory Fume Hoods	
1. Chemicals of low toxicity (TLV > 500 ppm)	75 – 80 lfm
e.g., acetone, ethanol, ethyl chloride, pentane, etc.	
2. Chemicals of moderate toxicity (TLV 10 – 500 ppm)	100 lfm
e.g., cyclohexane, methanol, turpentine, etc.	
3. Chemicals of high toxicity (TLV < 10 ppm)	125 – 150 lfm
e.g., aniline, carbon disulfide, ethanolamine, etc.	
Special Purpose Laboratory Fume Hoods	125 – 150 lfm
e.g., perchloric acid, radioisotope, etc.	

required to insure adequate protection for the worker. Table 10-1 is a suggested guideline for face velocities of laboratory fume hoods.

In a laboratory where workers spend most of their time working with chemicals, at least one hood for each two workers must be made available. The hoods must be large enough to provide each worker with at least 2.5 ft of working space at the face. In the event this amount of hood space cannot be provided, provisions for alternate types of local ventilation must be made available.

Laboratory workers must be trained in the safe use of hoods as follows:

(1) Hoods must be evaluated before use to ensure adequate face veloci- ties, typically 75 to 150 linear feet per minute (fpm), and the absence of excessive turbulence. Further, some continuous monitoring de- vice for adequate hood performance must be present and checked before each hood is used.

(2) The airflow pattern, and thus the performance of a hood, depends on such factors as placement of equipment in the hood; room drafts from open doors or windows; people walking by; or even the presence of the user in front of the hood. For example, the placement of equipment in the hood can have a dramatic effect on its perfor- mance. Moving an apparatus several inches back from the front edge into the hood can reduce the vapor concentration at the face by 90%.

(3) Except when adjustments of apparatus within the hood are being made, the hood must be kept closed. Keeping the face opening of the hood small will improve the overall performance of the hood.

(4) Hoods are not intended for storage of chemicals. Materials stored in hoods must be kept to a minimum. Stored chemicals must not block vents or alter airflow patterns. Chemicals must be moved from hoods to vented cabinets for storage.

(5)   Hoods must not be regarded as means for disposal of chemicals. Thus, apparatus used in hoods should be fitted with condensers, traps, or scrubbers to contain and collect waste solvents or toxic vapors and dusts.

(6)   If adequate general laboratory ventilation is certain and will be maintained when the hoods are not running, unused hoods can be turned off. However, if any doubt exists, or if toxic chemicals are present in the hood, the hood must be left on.

(7)   An emergency plan must be prepared for the event of ventilation failure, e.g., power failure, or other unexpected occurrence such as fire or explosion in the hood.

(8)   Hood fan and other local exhaust system components must be inspected at frequent intervals for corrosion, obstruction, and damage. A maintenance schedule including cleaning, lubrication, and adjustment must be established.

11 EMPLOYEE CHEMICAL EXPOSURE MONITORING

It is well established that exposures to toxic chemical air contaminants can be harmful, especially when the levels exceed permissible exposure limits. Laboratory work typically involves potential exposures to a variety of toxic chemical air contaminants under conditions such that the level of exposure could be excessive. Employee chemical exposure monitoring is done for a variety of reasons. The primary reasons are to identify and to quantify specific chemical contaminants present in the work environment. In the chemical laboratory, employee chemical exposure monitoring is simplified by two factors. First, the laboratory worker usually knows which contaminants are present in the laboratory from the nature of the reaction, plus a knowledge of the raw chemicals, end products, and wastes. Therefore, general identification of laboratory contaminants is rarely necessary and, as a rule, only quantification is required. Second, usually only a single contaminant of importance is present in the laboratory atmosphere and the absence of obvious interfering substances often permits great simplification of procedures. Nonetheless, one must be on guard continually to detect the presence of subtle and unsuspected interferences.

Other important reasons for employee chemical exposure monitoring in the laboratory include routine surveillance and evaluating compliance status with respect to various occupational health standards. In addition, air monitoring is conducted to determine exposures of laboratory workers in response to complaints and to evaluate the effectiveness of engineering controls, such as ventilation systems installed to minimize workers' exposure. Epidemiology of diseases of occupational origin and various other areas of research associated with chemical health are dependent on accurate evaluations of working and nonworking exposure to toxic chemical substances.

AIR CONTAMINANTS IN THE LABORATORY

The type of air contaminants occurring in the chemical laboratory may be divided into a few broad groups depending on physical characteristics:

Gases are fluids occupying the entire space of their enclosure and can be liquefied only by the combined effects of increased pressure and decreased temperature, e.g., carbon monoxide and hydrogen sulfide.

Vapors are the evaporation product of substances which are also liquid at normal temperatures, for example, methanol and water. Although thermodynamically gases and vapors behave similarly, the reason for making the distinction is because in many instances they are collected by different devices.

Particulate matter may be broadly divided into solids and liquids. In the solid group, there are three categories based on particle size and method of monitoring: dusts, fumes, and smoke.

Dusts are generated from solid inorganic or organic materials reduced in size by mechanical processes such as crushing, drilling, grinding, and pulverizing. The concern is for particles with an aerodynamic equivalent diameter of less than 10 μm because dusts remain suspended in the atmosphere for a significant period of time and are respirable.

Fumes are formed from solid materials by evaporation and condensation. Metals such as lead, when heated, produce a vapor which condenses in the atmosphere.

Smokes are products of incomplete combustion of organic materials and are characterized by optical density.

Liquid particles are produced by atomization or condensation from the gaseous state. Some of the terms used to describe liquid particles include *mists* and *fogs* but the distinction between these two terms has not been fully defined.

ENVIRONMENTAL FACTORS AFFECTING MONITORING

Evaluating the impact of airborne contaminants necessitates the accurate determination of the amount of the contaminant present in a unit volume of air. This value defines the concentration of the contaminant and is determined either directly from the airstream or following collection on a suitable medium. Calibrations are performed to establish the relationship between the instrument, between the instrument's response, and the airborne chemical contaminant concentration being measured. The reference standards used must be accurate and precise to produce well-characterized and reproducible calibrations.

Water vapor competes effectively with organic vapors for activated charcoal sites. High humidity, therefore, can significantly reduce the adsorption capacity of the charcoal that can be reduced by as much as 75% at relative humidities above 90%. The amount of reduction in adsorption capacity depends on the test compound and must be determined individually. Changes in barometric pressure and sampling temperature will also affect the collection efficiency and consequently the reported results. If the barometric pressure, elevation, or temperature conditions at the sampling site are substantially different from the calibration site, it would be necessary to recalibrate the sampling instruments at the sampling site where the same conditions are present. As a rule of thumb, if the barometric pressure difference is >3 kPa, elevation difference >300 m, and temperature difference >10°C, sampling site calibration of instruments will be deemed necessary.

GENERAL AIR MONITORING CRITERIA

An employee chemical exposure monitoring program must be designed to yield the specific information desired. In devising an employee chemical exposure monitoring program, it is essential to consider the following basic requirements:

(1) Any manipulation of air monitoring equipment in the field must be kept to a minimum.
(2) The air to be monitored must follow the shortest possible route to reach the collection medium.
(3) The air monitoring instrument must provide an acceptable efficiency of collection for the contaminants involved and the established efficiency must be maintained at a rate of air flow which provides sufficient sample for the intended analytical procedure for subsequent sample analysis.
(4) The collected air sample must be obtained in a chemical form that is stable during transport to the analytical laboratory. Consequently, any use of unstable or otherwise hazardous sampling media is to be avoided.

AIR MONITORING CONSIDERATIONS

The first step in conducting employee chemical exposure monitoring in any occupational environment is to become completely knowledgeable of the particular operation. The person evaluating the chemical laboratory must be aware of the reactions being run, the reagent chemicals used, and the end products, by-products, and chemical wastes encountered. In addition, he must know what personal protective measures are provided, how engineering controls are being used, and how many workers are exposed.

An experienced professional investigator, such as a trained industrial hygienist, often can evaluate quite accurately the magnitude of chemical and physical stresses associated with an operation on a qualitative walkthrough.

However, in order to document the actual airborne concentration of contaminants, it is necessary for the investigator to use air sampling devices and instrumentation. Regardless of the objectives of the employee exposure monitoring program, the investigator must take into consideration the following parameters in order to be able to implement the correct sampling strategy.

Area vs. Personnel Monitoring

The most common method of evaluating occupational exposure is to measure workroom contamination in the area of the workers in question. Measuring introduces a certain degree of uncertainty when evaluating the precise personal and specific exposure of the worker. It is, however, likely that many of these monitoring systems will continue in operation to monitor the effectiveness of engineering controls as well as administrative controls. Furthermore, area monitoring can generate valuable information on background exposure levels.

Ideally, one wishes to characterize the environment in the breathing zone of workers to evaluate individual exposure. A breathing zone is defined by OSHA to be a hemisphere forward of the shoulders with a radius of approximately 15 to 23 cm. Personal monitoring devices are especially useful for monitoring mobile employees who engage in a variety of operations involving different amounts of air contaminants of a diverse nature. For contaminants that are primary irritants or have permissible exposure limits, which include ceiling concentration, short-period personal samples are especially useful to define short-period maxima.

Volume of Air Sample

The volume of air sample to be collected is based upon the sensitivity of the analytical procedure, the estimated contaminant concentration, and the permissible exposure limit of the particular contaminant. Thus, the volume of air sampled may vary from a few liters, where the estimated contaminant concentration is high, to several cubic meters, where low contaminant concentrations are expected. Generally, the maximum amount of air to be monitored must be a balance between increased sensitivity of the analytical procedure and the economy of time.

Duration of Air Monitoring

Brief period samples are often referred to as *instantaneous* or *grab* samples. Longer period samples are termed *continuous* or *integrated* samples. Although there is no sharp dividing line between the two categories, grab samples are generally obtained over a period of less than 2 min. To determine peak concentrations of contaminant, grab samples are best. Integrated samples are taken for longer periods, ranging anywhere from a few minutes up to a full shift of 8 hr.

The prime objective in airborne chemical contaminant monitoring is to maintain an environment below the permissible exposure limit, which is called threshold limit value (TLV) by the American Conference of Governmental Industrial Hygienists (ACGIH). The TLVs refer to airborne concentrations of substances and represent conditions under which it is believed nearly all workers may be repeatedly exposed on a daily basis without adverse effect. Brief period samples include the threshold limit value-ceiling (TLV-C), the concentration which should not be exceeded even instantaneously, and the threshold limit value-short-term exposure limit (TLV-STEL), the maximal concentration to which workers can be exposed for a period up to 15 min continuously. The longest period sample is the threshold limit value-time-weighted average (TLV-TWA), the time-weighted concentration for a normal 8-hr workday for a 40-hr workweek. Threshold limit values have been incorporated into OSHA standards (29 CFR 1910.1000, Table Z-1 through Z-3), for evaluating the exposure of workers to airborne contaminants in the workplace.

Rate of Air Monitoring

Gas mixtures resist separation into components under the influence of centrifugal or inertial forces regardless of strength. Consequently, gas sampling presents no special problems with respect to sampling rate or velocity of entry into the sampling device. This is not the case for particulate matter sampling for particles < 5 μm in aerodynamic equivalent diameter, which is the most widely used definition of particle size in aerosol science. The definition is based on the way the particle behaves when airborne in a field of force. The formal definition is "the diameter of a unit density sphere which has the same settling velocity as the particle in question." The definition is based on behavior in a force field rather than on particle appearance. In addition to the aforementioned factors, other important factors to be considered in sampling rate are the total sampling time required, the dynamic characteristic of the sampling device, and the increase in sampling media resistance such as filters used in organic particulate matter collection.

Collection Efficiency

One of the most important factors in the collection of atmospheric contaminants is the efficiency of the monitoring device for the particular contaminant in question. In many cases, the efficiency need not be 100% as long as it is known and is constant over the range of concentrations being evaluated. It should, however, be above 90%. For many types of monitoring devices, the collection efficiency of the concentrating device must be measured.

Number of Air Samples

The number of samples to be collected depends to a great extent on the purpose of monitoring and the type of monitoring devices used. It is best accomplished, when analytical methods will permit, by allowing the worker to work a full-shift with a personal breathing zone monitoring device attached to the worker's body. Such a full-shift single sample demonstrates the worker's true average exposure level and the status of compliance or noncompliance with the exposure standard of the chemical contaminant. The major disadvantage of a full-shift single sample is that all environmental fluctuations of the chemical contaminant are averaged out during the monitoring period. Ideally, results of monitoring should reflect the true exposure levels relating to the worker's activities. In other words, the monitoring period should represent some identifiable period of time of the worker's exposure, such as a complete cycle of activity. This is particularly important in studying nonroutine or batch-type activities, which are characteristic of many laboratory operations. For this reason a continuous series of exposure measurements, either of equal or unequal time duration, should be obtained for the full duration of the desired time-average period. This type of monitoring is known as a full-shift consecutive sample. Other monitoring schemes include: (a) partial-shift single sample; (b) partial-shift consecutive samples; and (c) random grab samples.

MONITORING PROCEDURES AND DEVICES

Two basic methods exist for the collection of airborne contaminants. The first method involves the use of an air-moving device to obtain a definite volume of air at a known temperature and pressure. This method is known as active monitoring and, depending on the concentration and the analytical method used, the contaminant may be analyzed either with or without further concentration. The second method does not involve any air-moving device; rather, the air-monitoring device depends entirely on the phenomenon of diffusion of airborne contaminants to achieve trapping in a collection medium. This method is known as passive monitoring. Active monitoring can be applied to gases, and vapors, as well as particulate matter. Passive monitoring is currently used for gases and vapors only. Each of these monitoring methods is discussed in greater detail in the following:

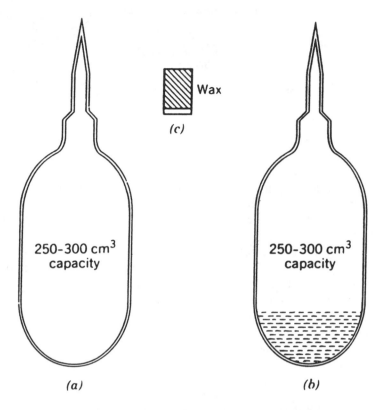

Figure 11-1   Evacuated flasks: (a) Empty evacuated flask; (b) Evacuated flask containing liquid absorbent; (c) Wax plug used to seal flask after sampling.

Active Air Monitoring

Active monitoring can involve either grab sampling or integrated sampling. In grab sampling, an actual sample of air is taken in a suitable container and the collected air sample is representative of the atmospheric conditions at the monitoring site at the time of monitoring. Numerous types of grab sampling devices are available. Included are evacuated flasks (Figure 11-1), gas or liquid displacement containers (Figure 11-2), flexible plastic containers, and hypodermic syringes.

Integrated sampling is used when the concentration of airborne contaminant is low, or the contaminant concentration fluctuates with time. It is also used when only the time-weighted average exposure value is desired. The contaminant in these cases is extracted from the air and concentrated by an absorbing solution or collected on an adsorbent. Absorbers vary in characteristic depending on the

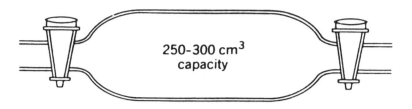

Figure 11-2   Gas or liquid displacement collector.

gas or vapor to be collected. Four basic types of absorbers are available: simple bubbling or gas-washing bottles, spiral and helical absorbers, fritted bubblers, and glass bead columns (Figure 11-3). The function of these absorbers is to provide sufficient contact between the contaminant in the air and the absorbing solution.

For nonreactive and insoluble vapors, adsorption is the method of choice. Commonly used adsorbents include activated charcoal, silica gel, and fluorisil. Adsorbing tubes, which are useful for personnel integrated sampling of most organic vapors, contain two interconnected chambers in series filled with the adsorbent. In an activated charcoal tube, for example, the first chamber contains 100 mg of charcoal; this chamber is separated from the backup section, containing 50 mg of charcoal, by a plastic foam plug (Figure 11-4). The contents of the two chambers are analyzed separately to determine if the charcoal in the first chamber has become saturated and lost an excessive amount of the sample to the second chamber. Monitoring results are invalid when the second chamber contains more than 10% of the amount collected in the first chamber.

A sampling train for active monitoring may consist of an air-moving device or source of suction; a flow regulator such as an orifice or a nozzle; a flowmeter to indicate the flowrate; a collection device such as an absorber, adsorbent, or a filter; a probe or sampling line; and a prefilter to remove any particulate matter that may interfere with sample collection or laboratory analysis.

Passive Air Monitoring

One of the more noteworthy developments in employee chemical exposure monitoring technology has been the availability of passive dosimeters for a broad list of vapors and gases. Such sampling devices are small, lightweight, and inexpensive. Passive dosimeters have no moving parts to break down. Further, they can be conveniently used unattended and depend solely on permeation or diffusion of gaseous contaminants to achieve trapping in a collection medium. Commercially available units usually require no calibration, as this is generally provided by the manufacturer.

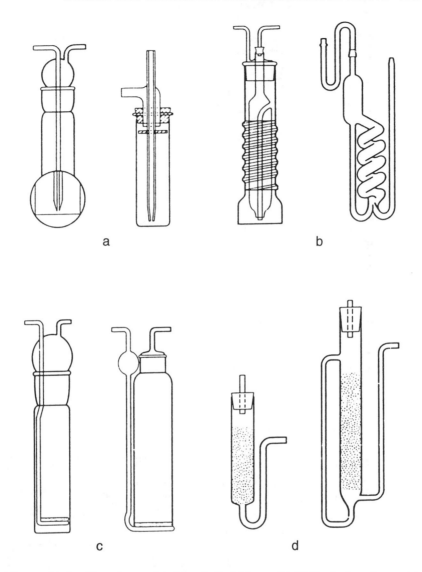

Figure 11-3   Absorbers: (a) Simple bubblers; (b) Helical absorbers; (c) Fritted bubblers; (d) Glass bead columns.

Permeation devices utilize a polymeric membrane of silicone rubber as a barrier to ambient atmosphere. Gaseous contaminants dissolve in the membrane and are transported through the membrane to a collection medium such as activated charcoal, silica gel, or ion exchange granules. Permeation across the membrane is controlled by the solubility of the vapor or gas in the membrane material and by diffusion of the dissolved molecules across the membrane under a concentration gradient.

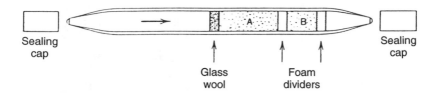

Sealing cap   Sealing cap   Glass wool   Foam dividers

Figure 11-4   Activated charcoal tube: (A) Sample portion; (B) Back-up portion.

Diffusion devices are provided with a porous barrier to minimize atmospheric turbulence. Molecules diffuse through the barrier to a stagnant air layer and then are collected on the adsorbent material. Specific information on sampling rates of the airborne contaminants is supplied by the manufacturers.

All the organic passive monitoring dosimeters use activated charcoal as the collection medium. Both permeation and diffusion devices are commercially available. They can be used to sample any organic compound that can be actively sampled by the charcoal tube method. It is important to note that the monitoring and analysis of organic passive dosimeters are affected by environmental factors such as atmospheric pressure, ambient temperature, relative humidity, and face velocity. The overall accuracy of passive dosimeters is ± 25% in the range of 0.5 to 2.0 times the environmental standard. It should be stressed that the manufacturer's recommendations on how and under what conditions the passive monitoring dosimeters may be used is to be strictly followed.

Colorimetric Indicator Tubes

Colorimetric indicator tubes provide conveniently compact direct reading devices for the detection and semiquantitative estimation of gases and vapors in atmospheric environments. At present, tubes for about 300 atmospheric contaminants are on the market. Indicator tubes have been widely advertised as being capable of use by unskilled personnel. Operating procedures are simple, rapid, and convenient. Yet, many limitations and potential errors are inherent in this method. The results may be dangerously misleading unless their use is supervised and findings are interpreted by a qualified industrial hygienist.

Colorimetric indicator tubes are glass tubes filled with solid granular material, such as silica gel or aluminum oxide, which has been impregnated with an appropriate chemical reagent. The ends of the glass tubes are sealed during manufacture. The use of indicator tubes is extremely simple. First, two sealed ends are broken open and the glass tube is placed in the manufacturer's holder, which is fitted with a calibrated squeeze bulb or piston pump. The recommended volume of air is then drawn through the tube by the operator. Adequate time must be allowed for each stroke. The observer then reads the concentration in the air by examining the exposed tube.

The earlier types of indicator tubes are provided with charts of color tints to be matched by the solid chemical in the indicating portion of the tube. Recent types of indicator tubes are based upon producing a variable length of stain on the indicator gel. A scale is usually printed directly on the tube and the result of monitoring can be obtained instantly. The range in the reading of results by different operators may be broad, since in many cases the end of a stain front may not be sharp.

An operator must take care to see that the pump valves and connectors are free of leaks. At periodic intervals, the flowrate of the apparatus must be checked and maintained within specifications. With most types of squeeze bulbs and piston pumps, the sampled air flowrate is high initially and low toward the end when the bulb or pump is almost filled. The claimed advantage is that the initially high rate gives a long stain and the final low rate sharpens the stain front. The general certification requirement for the accuracy of indicator tubes is ± 25% of the true value when tested at one to five times the TLV. The accuracy requirement is modified to ± 35% at one half the TLV. At best, indicator tubes may be regarded as only range-finding and approximate in nature. Furthermore, many of the indicator tubes are far from specific. An accurate knowledge of the possible interfering gases present is very important.

ANALYSIS OF AIR SAMPLES

The choice of which analytical method to use is frequently influenced by the sensitivities of the available approaches. In the field of industrial hygiene, heavy reliance has been placed on two techniques: atomic absorption spectrophotometry and gas chromatography. The inference does not constitute a claim that these two methods dwarf other analytical methods in importance. It does indicate the analytical versatility of these two popular techniques. Other analytical methods used in the Standards Completion Program of NIOSH include UV-VIS spectrophotometry, gravimetric analysis, fluorescence spectrophotometry, and high-pressure liquid chromatography, as well as others. Emerging analytical techniques that are beginning to find their way into the field of industrial hygiene include: inductively coupled plasma (ICP); optical emission spectrophotometry; electron spectroscopy for chemical analysis; X-ray fluorescence; ion chromatography; derivative spectroscopy; and Fourier transform infrared spectroscopy (FTIR).

INTERPRETATION OF RESULTS

Evaluating the impact of airborne chemical contaminants necessitates the accurate determination of the amount of the contaminant present in a unit volume of air by a qualified industrial hygienist. The task can be accomplished either directly from the airstream or following collection upon a suitable medium. A successful operation involves four areas that are of equal importance in the accurate determination of airborne contaminants: (1) calibration, (2) sampling, (3) analysis, and (4) data handling. The accuracy, and conversely the error, in the determination of the concentration of a chemical contaminant is a function of all four of these aspects of air sampling. To consider one without the other three often results in the incorrect or even improper determination of airborne concentrations.

12 MEDICAL EXAMINATION AND CONSULTATION

Laboratory personnel can be exposed to a variety of toxic chemicals, safety hazards, and physical hazards. Therefore, a sound medical examination and consultation program is essential to assess and monitor workers' fitness and health both prior to employment and during the course of work; to provide emergency and other treatment as needed; and to keep accurate records for future references. Information from medical examination and consultation programs may also be used to conduct future epidemiologic studies; to adjust claims, to provide evidence in litigation; and to report workers' health and medical conditions to federal, state, and local agencies, as required by law.

MEDICAL PROGRAM DEVELOPMENT

A medical examination and consultation program is the cornerstone of an effective health and safety management system. The program should be developed based on the specific needs and potential exposures of employees at the site, designed by an experienced occupational health consultant, and should contain the following components:

(1) Medical Examination

    (a) Pre-employment medical examination
    (b) Periodic medical examination and consultation
    (c) Termination medical examination

(2) Medical Treatment

    (a) Emergency medical treatment
    (b) Non-emergency medical treatment

117

(3)   Medical Record Keeping
(4)   Medical Program Review

Table 12-1 outlines a recommended medical examination program. Conditions and hazards of every laboratory very considerably. Thus, only general guidelines are given.

The success of any medical examination and consultation program depends both on management support and employee involvement. Management is encouraged to express its commitment by active participation. Prospective employees are urged to provide a complete, detailed occupational and medical history so that a baseline profile can be established. All employees must report any unusual physical or psychological conditions to the physician's attention, and report any suspected hazardous chemical exposure — regardless of the degree of exposure. Training programs must emphasize that even vague disturbances or apparently minor complaints such as headaches or skin rashes may be important. All employees should be assured of confidentiality of all medical information provided.

PRE-EMPLOYMENT MEDICAL EXAMINATION

The objectives of pre-employment medical examination are to determine a prospective employee's fitness for duty, including the ability to work while wearing personal protective equipment, and to provide baseline information for comparison with future medical data. To best accomplish these objectives, the pre-employment medical examination should focus on the following areas:

Medical and Occupational History

A medical and occupational history questionnaire is to be completed by the prospective employee. This questionnaire is to be reviewed by the occupational health physician before seeing the worker. In the examination room, the questionnaire is then discussed with the worker, with special attention to prior occupational exposures to chemical and physical hazards. Past illnesses and chronic diseases are reviewed. The worker's hypersensitivity to specific chemicals, cardiopulmonary condition, and personal lifestyle, such as tobacco and alcohol consumption, are reviewed.

Physical Examination

A comprehensive physical examination of all body systems and organs is to be conducted with a focus on the pulmonary, cardiovascular, and musculoskeletal

Table 12-1   Recommended Medical Surveillance Program

Component	Recommendations
Pre-Employment Medical Examination	1. Medical history. 2. Occupational history. 3. Physical examination. 4. Determination of fitness to work in personal protective equipment. 5. Freezing serum specimen for future testing.
Periodic Medical Examination	1. Annual update of medical history. 2. Annual update of occupational history. 3. Annual physical examination. 4. Testing based on examination results. 5. Testing based on specific exposures. 6. Testing based on job class and task.
Emergency Medical Treatment	1. Provide emergency care on site. 2. Develop liaison with local hospitals and medical specialists. 3. Arrange for decontamination of victims and personnel. 4. Arrange in advance for transport of victims. 5. Transfer medical records: give details of incident and medical history to next health care provider.
Non-Emergency Medical Treatment	1. Develop mechanism for non-emergency health care.
Medical Record Keeping	1. Maintain and provide access to medical records in accordance with federal, state, and local regulations. 2. Record and report occupational illnesses and injuries.
Medical Program Review	1. Review medical surveillance program periodically. 2. Review industrial hygiene and safety program periodically.
Termination Medical Examination	1. Final update of medical history. 2. Final update of occupational history. 3. Final physical examination. 4. Medical tests based on examination results.

systems. Conditions that could affect the worker's ability to work in the laboratory environment must be noted. In addition, note conditions that could affect the use of safety equipment, such as missing or arthritic fingers, facial scars, dentures, poor eyesight, or perforated eardrums.

Fitness to Work

Laboratory workers may be required to perform certain tasks while wearing personal protective equipment, such as a respirator and personal protective clothing, which may aggravate existing medial problems. To ensure that prospective employees are able to meet work requirements, individuals who are clearly unable to perform tasks based on the medical history and physical examination are to be disqualified.

Baseline Profile

Pre-employment examination can be used to establish baseline data to verify the efficacy of protective measures and to determine if exposures have adversely affected the worker. Baseline data include medical tests and biological monitoring. The latter may be useful for ascertaining pre-exposure levels of specific chemical substances to which the worker may be exposed. Table 12-2 are medical tests frequently prescribed by occupational health physicians. Pre-employment serum specimens may be frozen for future testing.

PERIODIC MEDICAL EXAMINATION AND CONSULTATION

Periodic medical examination and consultation must be developed and used in conjunction with pre-employment medical examination. Comparisons with the baseline profile is essential to detect trends that may mark early warning signs of adverse health effects, and thereby facilitate appropriate protective measures. The frequency and content of periodic medical examination and consultation will vary, depending on the nature of the work and exposures. A complete medical examination must be administered at least annually. More frequent examination and consultation may be necessary depending on the extent and nature of exposure, the type of chemicals involved, the duration of work assignment, and the worker's medical profile.

Periodic medical examination and consultation must include: updates of medical and occupational history, with a focus on changes in health status, illnesses, and exposure-related symptoms; physical examination; and additional

Table 12-2 Medical Tests

Function	Specimen	Sample Tests
Blood-forming function	Whole blood	Complete blood count (CBC) with differential and platelet evaluation including white cell count (WBC), red cell count (RBC), hemoglobin (HGB), packed cell volume or hematocrit (HCT), and other desired erythrocyte indices. Reticulocyte count may be appropriate if there is likelihood of exposure to hemolytic agents.
Kidney function	Blood serum	Blood urea nitrogen (BUN), creatinine, and uric acid.
Liver function/ General function	Blood serum	Albumin, bilirubin, globulin, and total protein.
Obstruction	Blood enzyme	Alkaline phosphatase.
Cell injury	Blood enzyme	Gamma glutamyl transpeptidase (GGTP), lactic dehydrogenase (LDH), serum glutamic oxaloacetic transaminase (SGOT), serum glutamic pyruvic transaminase (SGPT).
Multiple systems and organs	Urine	Analysis including color, appearance, specific gravity, pH, qualitative glucose, protein, bile, ketone bodies, and microscopic examination of centrifuged sediment.
	Stool	Occult blood.

medical tests based on available industrial hygiene and exposure monitoring information.

EMERGENCY MEDICAL TREATMENT

In planning for any potential emergencies, a wide range of actual and potential hazards must be considered. Emergency medical treatment must be included and integrated with the overall emergency planning program. Emergency medical treatment units must be available if a hazardous chemical incident occurs. Their primary objectives are to provide emergency medical treatment, decontamination, evacuation, and any other medical assistance to workers, victims, contractors, visitors, and any other personnel who may require emergency medical attention.

A trained team of emergency medical personnel must be established. The team will include members who have successfully completed a Red Cross or equivalent certified course in cardiopulmonary resuscitation (CPR), and first-aid training that emphasizes treatment of explosion and burn injuries, heat stress, and acute chemical toxicity. In addition, the team must include one or more emergency medical technicians (EMTs). A team of medical specialists headed by an occupational health physician must be available on an on-call basis for emergency consultation. In addition to having on site a standard first-aid kit or equivalent supplies, additional items such as antidotes for chemicals, decontamination solutions, potable water, emergency/deluge showers, and life support equipments must be available.

Standard operating procedures must be developed for emergency medical treatment of those injured victims who are also contaminated. During rescue operations, which include victim treatment and stabilization, emergency medical treatment personnel may have to determine if decontamination is necessary based on the type and severity of injury. Certain situations may exist where decontamination may aggravate subsequent health care or further delay a priority treatment. As a rule of thumb, if decontamination does not interfere with emergency medical care, proceed with it.

Plans must be made in advance for emergency evacuation to a nearby medical facility. Local emergency medical and transport personnel must be educated about the types of potential chemical hazards and anticipated medical problems involved in the evacuation plan. Hospitals must receive assistance in developing procedures for anticipated emergencies. The plan will help to protect hospital personnel and patients, and to minimize delays due to concerns about hospital safety or contamination.

NONEMERGENCY MEDICAL TREATMENT

Not all chemical substances exert immediate effects. The signs and symptoms of toxicity of many chemicals may not be present for 24 to 72 hr after exposure. In many instances, a hazardous chemical incident may be over, and the exposed workers may already be off duty and out of reach. All personnel operating at a chemical incident must be medically evaluated before being released. In addition, briefing for all chemical emergency response personnel on the signs and symptoms of the hazardous chemicals they have been exposed to must be provided. Instructions must also be provided on how and where to get immediate medical treatment if necessary.

Arrangements must be made for nonemergency medical care or consultation for emergency response personnel who are experiencing health effects resulting from an exposure to hazardous chemicals. A copy of the worker's medical records must be kept at the site and, when appropriate, at a nearby hospital under contract with the organization for medical care of its workers. All treating physicians should have access to these medical records.

MEDICAL RECORD KEEPING

Record keeping is an important element of any medical surveillance program because of the nature of the work and risks and because of the time interval between exposures and the appearance of their chronic effects. Medical records enable subsequent medical care providers to be informed about workers' previous and current exposures. Individual medical records must include all medical examinations and consultations completed, their purpose, and if they were a result of a specific exposure. A copy of the hazardous chemical incident report must also be maintained in the file. Any injuries sustained during line-of-duty operations must be noted and follow-up treatment, consultation, and personal exposure records must also be maintained.

Occupational Safety and Health Administration regulations mandate that, unless a specific standard provides a different time period, the employer must:

(1)   Maintain and preserve medical records on exposed workers for 30 years after they leave employment (29 CFR 1910.20).
(2)   Make available to workers, their authorized representatives, and authorized OSHA representatives, the results of medical testing and full medical records and analyses (29 CFR 1910.20).
(3)   Maintain records of occupational injuries and illnesses and post an annual summary report (29 CFR 1904).

MEDICAL PROGRAM REVIEW

Regular evaluation of the overall medical program is important to ensure its effectiveness. The following tasks must be reviewed on an annual basis:

(1)   Evaluate the effectiveness of medical testing in light of potential and confirmed exposures.
(2)   Add or delete specific medical tests as suggested by current industrial hygiene and environmental health data.
(3)   Ensure each accident or illness is promptly investigated to determine its cause and make necessary changes in health and safety procedures.
(4)   Review emergency and nonemergency medical protocols.

MEDICAL EXAMINATION UPON TERMINATION

To ensure the completion of a comprehensive medical profile, a medical examination must be given to all personnel when transferred from the laboratory or at termination of their employment. The medical examination administered immediately prior to termination of work or employment may be limited to obtaining an interval medical and occupational history of the period since the last full examination only if all three of the following conditions are met:

(1) The last full medical examination was within the last 6 months;
(2) No occupational exposure to hazardous chemicals has occurred since the last full medical examination; and
(3) No signs and symptoms associated with exposure have occurred since the last full medical examination.

If any of these three conditions listed are not met, a full medical examination is necessary at the termination of employment.

13   CHEMICAL EMERGENCY PLANNING

No matter how good the program, as long as any chemicals are used, there is the possibility that an accident will occur. On this assumption, it is advisable that chemical laboratories be prepared for emergencies. Generally, the first few minutes after the occurrence of a chemical emergency are the most critical. An immediate, quick, and effective response can have a tremendous effect on the situation. The correct response promotes situational control, and damages are minimized. A good chemical emergency plan can avert danger and prevent a catastrophe.

Accident prevention must always be the primary means of eliminating or minimizing damage, injuries to life, or material losses. However, pre-emergency planning is still a necessity. A chemical emergency plan which has been generated and for which equipment has been provided creates confidence even if the plan and the equipment are never used. From an economic standpoint, insurance rates may be reduced. Chemical emergencies vary in degree, therefore, emergency plans must be made accordingly for all potential chemical emergencies. Chemical emergency planning is the cornerstone of prevention. No single right way to write a chemical emergency plan exists. Just as laboratory conditions or the nature of operations constantly change, chemical emergency planning must be viewed as an ongoing process. A chemical emergency plan is a written document that must include but is not limited to the following:

CHEMICAL EMERGENCY OFFICER

The Chemical Emergency Officer is a key individual in the protection of health and safety at any hazardous chemical incidents. Each organizational structure is different. The individual responsibilities are in turn also different. In a small organization, one person may be immediately responsible for operation of safety, security, fire-fighting, and health services. In others, as is the case in large organizations, one person may be responsible for only one or two of select areas.

Any good chemical emergency plan must indicate who will have primary responsibility for directing operations once an emergency has occurred. Wherever possible, the Chemical Emergency Officer must be an experienced professional familiar with the particular problem that generated the emergency. A plant manager or a production supervisor may be knowledgeable of all plant operations, but if an accident occurs the plant manager or production supervisor must defer authority to the responding medical personnel, fire-fighters, or safety personnel.

The Chemical Emergency Officer will advise the response unit team leaders on all aspects of health and safety on site and will recommend stopping work if any operation threatens either worker or community health and safety. The Chemical Emergency Officer's responsibilities include but are not limited to the following:

(1) Select personnel protective apparel and emergency response equipment.
(2) Inspect personnel protective apparel and emergency response equipment periodically.
(3) Ensure that personnel protective apparel and emergency response equipment are properly stored and maintained.
(4) Confirm each response team member's capability to respond to emergencies.
(5) Monitor the response teams for signs of stress.
(6) Monitor on-site hazards and conditions.
(7) Participate in the development, implementation, and review of emergency plan.
(8) Know chemical emergency response procedures, evacuation routes, and the telephone numbers of outside resources if needed.
(9) Coordinate emergency medical care.
(10) Submit reports to federal, state, or local agencies as required.

POTENTIAL CHEMICAL EMERGENCIES

All potential chemical emergencies which may arise in the laboratory must be considered and listed in the chemical emergency plan. A chemical emergency is defined as any condition or action which could eventually lead to injury to life or damage to property. In preparing the written chemical emergency plan, all primary hazards should be listed first, e.g., a chemical fire, explosion, or sudden toxic chemical release, and others.

MAGNITUDE OF INJURY OR DAMAGE

The magnitude and type of injuries to lives, damage to property, or loss of materials that could occur in case of a chemical emergency must be considered. In other words, the number of workers who could be injured or even killed, instruments that could be damaged, or materials and production time that could be lost must be evaluated to determine the severity of a possible chemical emergency. In many cases, the outcome of such an analysis will dwarf the expenditures for any safety equipment.

CHAIN OF HAZARD EVALUATION

The initiating and contributory events that could lead to the primary chemical emergency should be determined. The causal chain of hazard evaluation will show different stages of the development of a chemical emergency and the safeguard or prevention method at each stage of hazard development can be determined.

PREVENTION METHOD

Safeguard or prevention method at each stage of hazard development is designed to minimize potential injury to lives, damage to properties, or loss of materials if control of the chain of events is lost. Such safeguards or preventive measures must be evaluated critically and periodically for adequacy. These safeguards or preventive measures may include but are not limited to: warning alarm system, automatic monitoring system, sprinkler system, emergency shut-down equipment, and many others.

CHEMICAL EMERGENCY RESPONSE PROCEDURE

A written procedure must be developed for rapid reaction to meet any anticipated chemical emergency. The available resources in personnel and equipment must be clearly indicated. All individuals and teams who will respond to chemical emergencies, their roles, responsibilities, and lines of authority must be clearly indicated. The procedure must consider the type of chemical emergency, its location, the unit to respond (assuming there is more than one), and the optimal means by which the problem could be combatted.

SITE MAP

A site map must be provided to each chemical emergency response unit and its personnel, indicating the best means of getting to the emergency area and alternative routes in case the primary route is blocked. Emergency routes must be examined as frequently as possible to determine if any obstructions may exist. Wherever the possibility of an obstruction may exist, the locations must be clearly marked to warn plant personnel against any unauthorized parking of vehicles, equipment. Traffic may otherwise be obstructed.

COMMUNICATION

A means must be provided for informing all emergency response personnel that a chemical emergency exists and the emergency response procedures must be instituted. The most effective means of communication must be established for response personnel, vehicles, and stations. The organization should also determine whether a special emergency number is to be established for reporting of emergencies or whether the telephone operator is to be called. This is especially true in the event of an emergency, since telephone operators are normally occupied with other lines, and additional channels of communication may be required. A special chemical emergency number to call to activate a chemical emergency response team is generally more advantageous, even if it is to the operator. The special chemical emergency number should be posted prominently on all phones.

A separate communication system is frequently necessary for emergency response personnel. Communication by telephone is generally dependable from the standpoint of mechanical or electronic reliability. Unfortunately, personnel often jam the telephone lines in an attempt to find out more about any emergency. In addition, a chemical emergency condition may affect or damage the telephone system. It may therefore be necessary to provide backup service through a radio network. Radio communication will allow chemical emergency response personnel to communicate directly with each other.

ALARM SYSTEM

Since different types of chemical emergencies may occur, several alarm systems must be provided. The alarms must be distinctive, different from other sounds, and last long enough so that the persons for whom they are intended will be properly alerted. They must also be instructive, informing personnel what and where the problem is. Needless to say, if a person collapses the unit responding will undoubtedly differ from the one responding to a chemical fire or explosion.

CONTROL OF UTILITY SERVICES

Emergency response personnel must know how to control utility services so that they do not add to the damage but are available where required in other parts of the organization. For example, it may be necessary to cut off the power running through downed high-voltage lines to ensure emergency personnel are not electrocuted or fires started. A ruptured natural gas line may have to be shut off. In addition, a circuit breaker may have to be closed to provide electricity for emergency equipment.

SITE SECURITY AND CONTROL

In a chemical emergency, the response unit team leader must know who is on-site and must be able to control the entry of site personnel into the hazardous areas to prevent additional injury to personnel. Only necessary rescue and response personnel may be allowed into the exclusion zone. In addition, safety zones and evacuation routes must be established immediately. The response team personnel must be able to determine rapidly the location of workers and the injured. An analysis must also be made rapidly to determine locations in which personnel will be safe or to which they can withdraw during an emergency.

EVACUATION

A severe chemical emergency, such as a chemical fire or explosion, may cut workers off from the normal exits. Routes to safety must be determined and analyzed for adequacy for the numbers of personnel by whom they would be used. Therefore, alternative routes for evacuating victims and endangered personnel must be established in advance, marked, and kept clear. OSHA standards require marking of routes, egresses, and exits so that they can be followed easily. The number of exits and egresses must be adequate for the time that would be available. They must also be adequate if one or more of them are blocked. Emergency lighting may be necessary if loss of the normal lighting system could throw the evacuation routes into darkness. Egress and exit signs must be lighted.

Safe distances for evacuation personnel can only be determined at the time of a chemical emergency. Safe distances are based on a combination of site- and incident-specific factors such as the quantity of chemicals released, the rate of release, and the wind speed and direction. A prior analysis must be made to determine locations in which personnel will be safe or to which they can withdraw during a chemical emergency.

One person, such as the laboratory supervisor, chemical hygiene officer, or other responsible person, must be designated to ensure that everyone leaves a specific area when it is to be vacated. This person must be trained to don appropriate personal protective equipment if necessary, to check the washrooms and other areas where workers might not be able to hear evacuation signals and to check for persons, especially handicapped workers, who might have trouble leaving expeditiously.

CHEMICAL EMERGENCY EQUIPMENT

In a chemical emergency, equipment may be necessary to rescue and treat victims, to protect emergency response personnel, and to mitigate hazardous conditions on-site. The various types of chemical emergency equipment needed to meet specific emergencies must be determined beforehand and purchased. All emergency equipment must be in working order, highly reliable, effective, and be available when an actual chemical emergency occurs.

The best locations for quick access to emergency equipment must be established. Further, the means of transportation and the personnel held accountable for the transportation must be determined. Equipment may either be stored near the sites of possible emergencies, or carried in by the response personnel. If the emergency equipment is to be stored, its storage site must be located in close proximity to where the equipment might be needed. Storage sites must not be so located that the condition creating the emergency prevents reaching it, or rendering its use ineffective. Storage units must be easily accessible and marked for quick identification. Organization directives must specify unblocked access to the storage units, prohibit removal of any material except by authorized personnel, and conduct periodic inspection of all emergency equipment. A log must be established to indicate the date and by whom the inspection was made and to note any discrepancies.

RESCUE

In a chemical emergency, the possibility exists that persons involved may not be able to escape using their own resources. Rescue provisions must therefore be provided. Rescue devices must be foolproof in an emergency, require a minimum amount of effort to operate, and be easily operated when only minimal instructions are provided. The instructions must be marked, easy to recognize, and easy to understand by a person under stress. The presence of effective devices with suitable markings can mean the difference between a successful rescue attempt and failure. Periodic tests must be made to ensure that escape and rescue devices work properly when used in accordance with the instructions provided.

DECONTAMINATION

In some chemical emergency situations, decontamination may be an essential part of life-saving first aid. In some other situations, decontamination may aggravate the injury or delay life-saving efforts. The decision whether to decontaminate a victim must be based on the type and severity of the injury and the nature of the chemical contaminants. If decontamination does not interfere with essential treatment, proceed with it. If decontamination cannot be performed, the victim must be wrapped in blankets, plastic or rubber sheets, to reduce contamination of other personnel. Off-site emergency medical personnel must be alerted to the potential contamination, or to specific decontamination procedure.

EMERGENCY MEDICAL TREATMENT

In emergencies, medical treatment may range from bandaging of minor cuts and abrasions to life-saving techniques. In many cases, essential medical help may not be immediately available. For this reason, it is vital to train on-site emergency response personnel in on-the-spot treatment techniques, to establish and maintain telephone contact with medical experts, and to establish liaisons with local hospitals and ambulance services.

TRAINING FOR CHEMICAL EMERGENCIES

All personnel within an organization using chemicals must receive instructions on emergencies, commensurate with their work and to the potential hazards to which they are exposed, and on what to do in a chemical emergency. Training must be supplemented by reminders such as awareness posters. Laboratory supervisors must ensure that they as well as their workers are familiar with such instructions. Each person can provide vital assistance if he or she is on the spot when a chemical emergency arises and can take immediate corrective action. It is also important for each worker to know when it would be more advantageous to obtain more expert assistance first rather than to waste time trying to combat the problem.

The importance of adequate and frequent training for chemical emergency response and rescue personnel can never be overemphasized. All these personnel must become proficient in carrying out the prescribed chemical emergency procedures. Simulated chemical emergencies and frequent drills will help increase proficiency. Investigations of many serious accidents have revealed that frequently personnel have either died because of a lack of proficiency in the use of emergency equipment/devices or because of failure to follow established procedures.

FOLLOW-UP AND REVIEW

Immediately following a hazardous chemical emergency, the Chemical Emergency Officer must conduct an investigation and document the incident. The follow-up document is especially important if the incident resulted in personal injury or property damage to the surrounding environment. In addition, there may be reporting requirements to the Department of Transportation, the Environmental Protection Agency, or the Occupational Safety and Health Administration. Before normal activities are resumed, all personnel must be fully prepared and equipped to handle another chemical emergency. All emergency equipment must be refurbished, repaired, and ready for use again. The site safety and chemical emergency plan must be reviewed and revised per new site conditions and lessons learned from the last chemical emergency.

14 CHEMICAL EMERGENCY RESPONSE PROCEDURES

The very nature of the work with hazardous chemicals makes emergencies a continual possibility, regardless of their infrequency. Hazardous chemical disasters strike quickly and unexpectedly and require immediate response. A hazardous chemical emergency may be as limited as a worker breaking a glass bottle of mineral acid, or as extensive as an explosion spreading toxic chemical fumes throughout the community. Chemical mishaps are always characterized by their potential for complexity. Toxic and hazardous chemicals may react with each other, resulting in synergistic effects, or they may potentiate each other. Rescue personnel attempting to remove injured fellow workers may themselves become victims. Advance planning, anticipation of various disaster scenarios, thorough preparation for contingencies, and frequent disaster drills are essential elements necessary to protect the health and safety of the chemical emergency responders and the community.

Decisive action is immediately required when a hazardous chemical incident occurs. Remember instantaneous decisions may have far-reaching, long-term consequences. Delays of minutes can create irreparable damages or life-threatening situations. Personnel must be ready to be dispatched at a moment's notice and be ready to respond and rescue. Emergency supplies and equipment must be stocked in sufficient quantities, readily available, and in good working order.

Chemical emergency response personnel may be deployed in a variety of ways. The nature and scope of the emergency, the size of the site, and the number of personnel involved determine the composition of the response teams and the capabilities of their deployment. Emergency responders may be a couple of individuals or groups of varying sizes which constitute a team. Further, several teams may interact with each other. Although deployment is determined on a site-by-site basis, the organizational structure must show a clear chain-of-command, each individual must know his role, and the chain-of-command must

133

be flexible enough to handle multiple emergencies. Immediate, informed response is essential in a disaster, as is the comprehensive training of personnel.

PERSONNEL PROTECTION

Personnel protection is the key element in a safe emergency response to a hazardous chemical incident. The individual components of protective apparel, respirator, and other equipment must be assembled into a full protective ensemble that both protects the responder and minimizes the hazards of the ensemble itself. When approaching an unknown or undefined hazardous chemical incident, full chemical protection gear must be worn to protect the emergency responder from adverse exposures until the actual conditions can be determined. Once the site situation is determined, an adequate level of protective apparel and equipment can be established. The overall level of protection must be reevaluated periodically as the amount of information about the site increases. Emergency response personnel should be able to upgrade or downgrade their level of protection with concurrence and approval of the emergency response commander.

The Environmental Protection Agency (EPA) has its own system for classifying protective clothing. Appendix C lists ensemble components based on the EPA Levels of Protection: levels A, B, C, and D. These lists of personal protective equipment can be used as a starting point for ensemble creation. However, each ensemble must be tailored to the specific situation in order to provide the most appropriate level of protection. It should be realized that typical turnout gear or structural fire-fighting clothing is not considered full chemical-resistant protective clothing. Rather, it provides wearer protection similar to the EPA's Level B or Level C.

Level A Protection

Level A provides maximum protection against vapors, gases, mists, and even dusts. The garment required to provide this protection is called a fully encapsulating chemical-resistant suit. This means that the ensemble envelopes the wearer totally, including the self-contained breathing apparatus (SCBA). The garment material is quite heavy and provides longer protection before breakthrough by the chemicals. The chemical emergency planning group must carefully analyze the various categories of chemicals they wish to provide protection against and then study which fabric provides the best spectrum of protection against those chemicals.

Level B Protection

Level B protection is worn when a high level of respiratory protection is needed but a lower level of skin protection is required than at level A. The level B chemical splash suit is nonencapsulating. The SCBA is worn on the outside. This ensemble is used where the chemicals involved do not generate vapors of a life-threatening nature. If the chemical substance is flammable and ignition sources are a real threat, fire-fighting turnout coats and pants or a flash fire protection garment should be worn over the level B suit. The chemical splash suit is usually two-piece, similar to the heavy rainwear worn by highway maintenance personnel.

Level C Protection

Level C protection differs from level B in the equipment for respiratory protection. The same level of chemical-resistant clothing is used for skin protection. Level C allows the use of non-fire service respirators.

Level D Protection

Level D protection is primarily the basic work uniform. The atmosphere will not contain a known hazard and the work functions will preclude splashes, immersion, or the potential for unexpected inhalation of or contact with hazardous levels of any chemicals.

HAZARD AREA CONTROL

In a hazardous chemical incident, the response unit team leader must control the entry of personnel into the hazard area to prevent any exposure. Only necessary and trained personnel can be allowed into the hazard area to perform their functions. Concerning the limited resources of first responders, they must direct their attention to security and control. In other words, they must establish the following zones of hazard as soon as possible:

Exclusion Zone

The exclusion zone is also referred to as *restricted* zone. This is the immediate area surrounding the hazardous chemical incident itself and the area where probability of exposure to the hazardous chemicals is highest. Entry is to be restricted to only those with proper level of personal protection and training. It is becoming extremely popular to identify a restricted zone by using red barrier tape which establishes a contamination perimeter. Hence, the restricted zone is also referred to as a *red* zone or *hot* zone.

Limited Access Zone

The limited access zone is also referred to as *contamination reduction* zone. This is a large geographical area surrounding the restricted zone. First responders should limit area access to necessary personnel only. Unnecessary personnel, such as the media or police and fire-fighters without proper personal protection gear, should not be allowed in. It is common to identify a limited access zone using yellow barrier tape to establish a safety perimeter. Hence, the limited access zone is also referred to as *yellow* zone or *warm* zone.

Support Zone

The support zone is the area encompassing the limited access zone. It is considered a safe area where emergency response personnel can move freely. The command post and staging area should be established in this zone, along with any other required logistical operations. The spread of the hazardous chemical to this zone is unlikely unless there is major deterioration of the situation. The boundary of this zone is to be established immediately by the first responders to prohibit bystanders, the media, and other nonessential personnel from becoming exposed to the possible dangers of the situation. Rope or traffic cones can be used to identify the boundary establishing an isolation perimeter. The support zone is sometimes referred to as *green* zone or *cold* zone.

Refuges are on-site safety stations and should be set up only for essential needs such as short rest breaks or for emergency responder strategy meetings. The refuges should be located in a relatively safe but not necessarily "clean" area. Refuges are never to be used for activities such as eating or drinking. Additional refuges can be set up in the support zone to provide for emergency needs such as: first aid for injured personnel; clean, dry clothing; washwater for exposed victims; and communications with the command post.

Detailed information about the site is essential for effective response to any hazardous chemical incident. A site map will not only serve as a graphic record of the hazard, but can also be used to define zones of activity, determine evacuation routes, and identify various stations and activities. Since a sudden worsening of the situation may cut personnel off from the normal exit near the command post, alternate routes for evacuating personnel and victims should be established. Alternative evacuation routes should be directed from the restricted

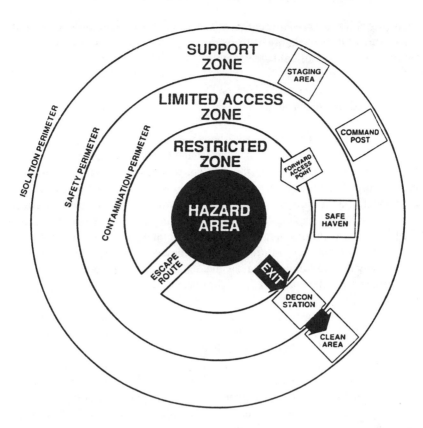

Figure 14-1   Hazard area control.

zone through an upwind limited access zone to the support zone and from the support zone to an off-site location in case conditions necessitate a general site evacuation. The concept of hazard area control is summarized in Figure 14-1.

CHEMICAL HAZARD EXPOSURE ASSESSMENT

Airborne toxic contaminants can present a significant hazard to chemical emergency response personnel. Thus, identification and quantification of these toxic contaminants is essential in the determination of the types and levels of personal protection, the need for specific medical and biological monitoring, and the assessment of the potential health effects of exposure.

Initial airborne toxic hazard assessment for identification is often qualitative, i.e., the contaminant, or the class to which it belongs, is demonstrated to be present but the determination of its concentration or quantification must involve more elaborate testing. Two principal approaches are available for the identification and/or quantification of airborne chemical contaminants: (a) the on-site use of direct-reading instruments; and (b) the use of methods requiring laboratory analysis of collected samples.

Direct-Reading Instruments

Direct-reading instruments were developed as early warning devices for use in industrial settings, where a leak or an accident could release a high concentration of a known chemical into the ambient atmosphere. Today, some direct-reading instruments can detect contaminants in concentrations down to one part contaminant per million parts of air (ppm), and provide instant information at the time of sampling, enabling rapid decision-making. It is imperative that direct-reading instruments are operated, and their data interpreted, by qualified industrial hygienists familiar with the particular device's operating principles and limitations. At hazardous chemical incidents, where unknown and multiple contaminants are the rule rather than the exception, instrument readings should be interpreted conservatively.

Direct-reading instruments are of two general groups. The first group consists of colorimetric indicator tubes as discussed in Chapter 11. The second group is comprised of electronic circuitry capable of: sampling a volume of air; performing qualitative and/or quantitative analysis internally; and displaying the results immediately on a dial, illuminated digital display panel, tape printout, or strip chart recorder. Direct-reading instruments of the second group commonly carried by first responders include organic vapor monitors, combustible gas indicators, oxygen meters, and others.

The rapid development of portable and sensitive electronic direct-reading instruments for evaluating ambient atmospheres has largely been due to the borrowing of air sampling and analysis technology previously developed in the disciplines of radiation protection and air pollution control. The class of direct-reading instruments incorporates electronic sensors utilizing infrared and ultraviolet radiation, flame and photoionization, and chemiluminescence capable of detecting and measuring airborne concentrations of gases and vapors in a matter of seconds.

Many gases and vapors have characteristic infrared spectra that absorb infrared radiation over a spectrum of wavelengths in a manner that can be converted into characteristic graphs. Such characteristics can be used to detect the presence of air contaminants and to determine their concentrations in air. Recent advances in technology have reduced the power requirements for infrared analyzers formerly requiring permanent installation. Today, these highly sensitive devices can be housed in a portable instrument design.

Another type of direct-reading instruments using the principle of spectroscope is the ultraviolet photoionization detector. Ultraviolet photoionization detectors are used to read the total concentrations of many organic and some inorganic gases and vapors. The unit operates by the principle of using ultraviolet radiation to ionize molecules which produce a current proportional to the number of ions. Some of the limitations associated with these detectors include: methane is not detected; the detector is affected by high humidity; and extensive calibration is required.

A portable combustible gas meter, commonly called an explosimeter, is available to monitor atmospheres for concentrations of combustible or flammable gases and vapors. This instrument is calibrated in percent lower explosive limit (% LEL) for a single gas, usually methane or hexane. Its accuracy depends upon the differences between the calibration gas and the gas being sampled. Manufacturers of combustible gas meters provide calibration curves for the common combustible gases. An important feature of these meters is an audible alarm which sounds when vapor concentrations reach 25% of the LEL. A second important feature is a self-operating sampling system which automatically samples the air mixture. Therefore, these meters are generally sensitive enough to detect the presence of a broad range of flammable gases and vapors before they reach their lower flammable limit in air. These instruments are well suited for testing the atmosphere for the presence of solvent vapors, volatile vapors, and natural and manufactured gases. A red alarm light and a horn inside the unit are activated when the concentration of flammable gases reaches the preset limit, providing both visual and audio warnings of a dangerous concentration.

Air normally contains about 21% oxygen by volume. Although oxygen does not have a specific threshold limit value, its level in industrial air must often be measured, particularly in enclosed areas where combustible or other processes may use up the available oxygen. OSHA regulation (29 CFR 1910.94) states that if air contains less than 19.5% it is oxygen-deficient. However, the OSHA Maritime Standard (29 CFR 1915.12) considers air containing less than 16.5% as oxygen-deficient. Both direct-reading and indirect-reading instruments are available to sample breathing air for oxygen content. Direct-reading instruments are small, lightweight, and easy to use, and include those based on the coulometric principle and on colorimetric and paramagnetic analysis.

Flame ionization provides a sensitive method of detection of the total concentrations of various hydrocarbons in air. The flame ionization detector works under the principle that gases and vapors are ionized in a flame and a current is produced in proportion to the number of carbon atoms present. A loop of platinum held above the flame serves as the collector electrode. The current carried across the electrode gap is directly proportional to the number of ions produced. This detector responds to all organic compounds except formic acid, and its response is greatest with hydrocarbons.

Laboratory Analysis

Direct-reading instruments are available for only a few specific substances and are rarely sensitive enough to measure the minute quantities of specific hazardous chemicals which may induce health effects. Thus, to detect relatively low-level concentrations of specific hazardous chemicals, it may be necessary to collect air samples and have them analyzed in a laboratory. Selection of the sampling media, sampling time, and sample volume has previously been discussed in Chapter 11. A major disadvantage of long-term air sampling is the time required to obtain data. The time lag between air sampling and obtaining the analytical results may be hours. Today, mobile laboratories are available and may be brought on-site to analyze hazardous chemicals. A mobile laboratory is generally a trailer truck housing analytical instruments capable of rapidly analyzing chemicals by a variety of techniques. Typical instruments include gas chromatographs, spectrophotometers, chemiluminescence, and other techniques. When not in use in the mobile laboratory, these devices can be relocated to fixed-base facilities.

LEVELS OF EXPOSURE

In all probability, the concentration of any released hazardous chemical will be greatest at the site of incident and decrease as the distance from it increases. As a first step, air sampling must be conducted to identify any condition that may be immediately dangerous to life and health (IDLH). In addition, other conditions such as flammable or explosive atmospheres, oxygen-deficient environments, and highly toxic levels of airborne concentrations should also be identified.

IDLH Concentrations

Exposure concentrations which are IDLH have been established by the NIOSH/OSHA Standard Completion Program as a guideline for selecting respirators for some chemicals. The definition of IDLH concentration varies and depends on the agency or organization. The *NIOSH Pocket Guide to Chemical Hazards* defines IDLH concentration as the maximum concentration from which in the event of respirator failure, one could escape within 30 min without a respirator and without any escape-impairing symptoms or any irreversible health effects. The Mine Safety and Health Administration (MSHA) standard defines IDLH conditions as those which pose an immediate threat to life or health or that pose an immediate threat of severe exposure to contaminants such as radioactive materials which are likely to have adverse cumulative or delayed effects on health. The American National Standards Institute (ANSI) defines IDLH as any atmosphere which poses an immediate irreversible debilitating effect on health. Regardless of the definition, all IDLH concentrations indicate those exposure levels of toxic substances from which escape is possible without irreversible harm should a worker's respiratory protective device fail. At hazardous chemical incidents, IDLH concentrations should be assumed to represent concentrations above which only workers wearing respirators providing the maximum protection such as a positive-pressure, full-facepiece, SCBA or a combination positive-pressure, full-facepiece, supplied-air respirator with positive-pressure SCBA are permitted. Specific IDLH values for some chemical substances can be found in the *NIOSH Pocket Guide to Chemical Hazards*.

Threshold Limit Values

Threshold limit values (TLVs) refer to airborne concentrations of substances and represent conditions under which it is believed that nearly all workers may be repeatedly exposed day after day without adverse effect. Discussed in Chapter 11, these values have been derived for a number of chemical substances and can be found in *Threshold Limit Values and Biological Exposure Indices* which is published annually by the American Conference of Governmental Industrial Hygienists (ACGIH).

"Skin" Notation

Skin absorption of chemicals can contribute to the overall exposure. Some chemical substances may carry the "skin" notation, indicating potential contribution to the overall exposure by the cutaneous route including the mucous membranes and the eye, either by airborne, or more specifically by direct skin contact with the chemical. Vehicles can also alter skin absorption. Chemicals having a "skin" notation and a low TLV may present a problem at high airborne concentrations, particularly if a significant area of the skin is exposed for a long period of time. This attention-calling notation is intended to suggest appropriate personal protective measures for the prevention of cutaneous absorption. Biological monitoring should be considered to determine the relative contribution of dermal exposure to the total dose.

ROLE OF THE FIRST RESPONDER

Today, the phrase "hazardous chemical incident" has a whole new meaning in emergency response. There are new rules which must be followed, particularly by first responders. The first responder must share whatever information is collected at the scene with his team, with other responding units, and with local, state, and federal agencies if they are involved or required by law.

The first responder must be prudent to identify the need to pull back, if necessary, or even withdraw completely. Whatever decisions initially made by the first responder, it will set the tone for all future decisions and actions. If the activities of the first responder are well thought out, logical, and performed in a professional manner, success should be achieved without undue danger. Illogical action at any level of chemical emergency response is risky and should be avoided. Coordinated efforts offer substantial margins both in safety and operational efficiency.

Anything important to the response noted by the first responder should be reported to the communications center for relay to other incoming units and to advisory consultants as they are contacted. Observed conditions should be compared to reported conditions and a determination made as to the relative change over the time period. Whether a chemical emergency appears to be stable or has changed greatly since the first report will affect the selection of strategy. Upon arrival at the scene, the first responder should immediately set the chemical emergency plan in motion.

INITIAL RESPONSE ACTION

Upon arrival to the scene of a hazardous chemical incident, the first responder must make some quick decisions and take immediate action. An initial survey of the hazard area must be made immediately to determine the location of those who may be threatened or affected; the presence of fire, smoke, or fumes; and the overall condition of the environment. Wear the maximum degree of personal protective clothing and equipment practicable. Plan an escape route before entering the hazard area. Rescue the injured if possible.

Chemical Hazard Identification

The first priority when a hazardous chemical incident occurs is the prevention of injuries and fatalities to people. The second priority is the prevention of damage to property and the environment. To handle a hazardous chemical incident in the safest manner, it is absolutely necessary to know the chemicals involved. Actual response method and cleanup technique used may vary depending on local conditions and the environment. It is possible that the scene of a hazardous chemical incident will represent such a high degree of hazard that the only safe course is to protect the perimeter and evacuate those who may become exposed to the toxic fumes which may exist with or without the presence of fire.

Hazard Area Control

The first responder may find, upon arrival, little can be done to eradicate or lessen the hazard. In such a case, it is best to do nothing and proceed no further. Instead, the first responders must isolate, deny entry, and protect the chemical hazard area until proper resources are assembled to mitigate the situation. To protect the scene, the first responder must immediately identify an isolation perimeter around the incident, and establish a support zone. Fire and police personnel should not be allowed beyond this zone unless they have the appropriate personal protective apparel and equipment. All nonessential personnel and news media people must be kept beyond the isolation perimeter. The second zone is the limited access zone and it is identified by yellow barrier tape. Identifying this zone reduces the chance of hazardous materials spread by personnel contamination. The first responder should go to the same lengths to protect the scene as would be done to protect the entire responding team.

Evacuation and Safety

Evacuation of people downwind from a hazardous chemical incident is an extremely difficult task. The EPA has conducted a study on evacuation of the public. The study concluded that the biggest problem is getting people to move at all. Once a toxic gas or vapor is released, staying or going indoors to a "safe haven" may provide substantial protection against instantaneous or short to medium duration releases. People seeking a safe haven may have a better chance of survival than someone outdoors.

Chain of Command

The management system at a hazardous chemical incident should clearly identify who is in command. The person in control sets up a chain of command, according to the procedures in the chemical emergency response plan. With command established, information can be gathered and the need for additional manpower, equipment, and support resources can be identified. All pertinent aspects regarding the hazardous chemical incident must be communicated to the person in command. All directives given by the on-scene commander to personnel must be clearly stated, and an understanding of these directives confirmed.

Site Characterization

Site characterization is the process of data gathering and assessment. It provides the information needed to identify site hazards and select emergency responder protection methods. Accurate, detailed, and comprehensive information concerning site characteristics will enable the response personnel to tailor protective measures to the site's actual hazards. The person with primary responsibility for site characterization and assessment is the response unit team leader. Site characterization generally proceeds in three phases: off-site survey, on-site survey, and air monitoring. The information obtained at each phase is used to assess hazards and to develop or modify a site safety plan for the next phase of characterization.

Off-Site Survey

Prior to initial site entry, information from off-site interviews, visual observation, records research, and air monitoring at the perimeter must be obtained. This information is then used to evaluate the potential hazards, particularly inhalation hazards which may be IDLH as well as other dangerous conditions, and to develop a site safety plan for initial entry which specifies protective equipment and other safety controls.

On-Site Survey

The purpose of this phase of site characterization is to verify and supplement information from the off-site survey. After careful evaluation of probable conditions based on off-site data, priorities must be established for hazard assessment during the on-site survey. The composition of the entry team depends on the site characterization, but should always consist of at least four people: two will enter the site and the other two will serve as outside support people. Upon entering the site, responders should first monitor for IDLH and other dangerous conditions. If these hazards do not exist, air monitoring and visual inspection of the site may proceed. Information from the on-site survey is used to assess the safety of the site for cleanup activities and to modify the site safety plan for the next phase of activity. Any physical hazards present must be controlled or removed before cleanup.

Site Air Monitoring

During cleanup, air monitoring must be performed in order to provide ongoing information about site conditions. Site monitoring includes evaluation of any changes in site conditions or work activities affecting the safety of responders. It is important to recognize that site characterization is an ongoing process. In addition to the formal information gathering which takes place during the three phases of site characterization, emergency response personnel must be constantly alert for new information about site conditions.

Documentation

Site characterization provides the database upon which subsequent actions and decisions are based. Therefore, good documentation is essential. All information must be carefully recorded using pre-established means such as logbooks, graphs, photographs, sample labels, and chain of custody forms. Control of all documents is necessary to ensure project accountability. In addition, reporting of hazardous chemical incidents may be required by law.

CONTAINMENT TECHNIQUES

Containment is the isolation or confinement of free-flowing material to minimize the area of the spill and to prevent contamination of other uncontaminated areas. The feasibility of containment of a spill of hazardous chemicals is one of the preliminary evaluation decisions to be made by the first responder. The method for containment of a spill will vary, depending upon the nature of the chemical spilled, the quantity, and the location of the spill.

Liquid Spills

Many liquid chemical spills pose a significant vapor hazard. Free-flowing running spills must be contained or absorbed immediately to avoid spread of the chemical. In order to contain a liquid spill, the following points must be considered: the topography of the area; potential runoff contamination of the natural waters, sewers, or even the soil.

Absorbents are commonly used for immobilization and collection of liquid hazardous chemical spills. They are of two distinct groups: universal absorbents and oil absorbents.

Universal absorbents absorb all liquids including water and are normally composed of silicates, clays, organic material, or various types of fabric. Universal absorbents range in form from rectangular pillows to sheets, rolls, and bags of loose granular material. Table 14-1 lists some commonly used universal absorbents.

Oil absorbents are used in cleanup operations to absorb oils and other petroleum-based hydrocarbons. Oil absorbents that actively repel water and normally float on its surface are indispensable for absorbing floating hydrocarbon spills from water courses. Oil absorbents are marketed in a variety of forms such as rolls, sheets, sweeps, pillows, particulates, and booms.

Adsorption is the physical entrapment of organic contaminants in the granular pore structure of solid material. The most common type of adsorbent is

Table 14-1   Commonly Used Absorbents

Absorbent (Supplier)	Comments
3M Brand Universal Sorbents (3M Corporation)	Available in sheets or rolls.
Conwed Universal® Absorbents (Conwed Corporation)	Available in pads or rolls.
HAZMAT PIG (New Pig Corporation)	Available in tubular socks. Acid resistant.
Hazorb® Pillows (Occidental Chemical Corp.)	Available in pillows or in loose granules. Absorbs virtually all liquids. (Acids/alkalis/solvents).
Safe Step Absorbent (Andesite of California, Inc.)	Available in granular form.
Solid-A-Sorb (Eagle Picher, Inc.)	Available in bags.
Speedi-Dry (Oil Dry Corporation)	Available in bags. Useful for diking.
Toxi-Dry (Michael Woods Products, Inc.)	Available in bags. Not compatible with strong acids or oxidizers.

granular activated charcoal. For years, activated charcoal has been used in the air and water pollution control field for removal of organic contaminants from air and water streams.

Ordinary absorbents and adsorbents may entrap organic molecules in their solid structures, but many are ineffective in vapor suppression. Another dimension is added in containment if a flammable liquid is spilled. Fire-fighters are acutely aware of the potential danger of flammable liquid spills. One of the commonly used fire-fighting techniques is the application of high and low expansion foams to suppress the vapor of flammable liquids. It is important to note that free-flowing running flammable liquid spills must be contained before using the foam. If the spill is already on fire, then the fire must first be extinguished before the vapor suppression foam is applied. Appendix D contains a partial list of chemicals that have been successfully tested using the HAZMAT® and Universal® vapor suppression, fire-fighting foams.

Field experience demonstrates some chemical vapors can be scrubbed or dispersed in the atmosphere by using hand-held water fog lines. For years, major suppliers of anhydrous ammonia have held training sessions with local and industrial fire brigades to train them in the techniques of water fog absorption and dispersion of ammonia vapors. Fog line dispersion works best on water-soluble chemicals. The effectiveness of using fog lines for vapor reduction is more often attributed to the lofting and mixing of the vapors with air, thereby increasing dispersion of the vapor cloud.

Solid Material Spills

Solid material spills in the laboratory do not present any serious cleanup problems. If a spill occurs outdoors on land, the solid material spill must be isolated immediately by covering the spill with plastic sheeting or other compatible covering to prevent dispersion by wind, rain, or other forms of precipitation. If the spilled solid material is also flammable, care must be taken to isolate the material from ignition sources.

LEAK-STOPPING DEVICES

The mechanical means to stem the flow effectively from a container or a stationary source of a hazardous chemical release may involve the use of one or more of the following devices or methods:

Repositioning the Leaking Container

In order to stem the flow of a liquid from a drum spill, frequently all that is required is to manually reposition the container so that the leak is on the top. After repositioning the drum, the spilled liquid chemical must be cleaned up prior to patching the drum.

Epoxy Cements

Epoxy cements are generally paste-like or putty-like in texture and come in a "fast cure" two-part blend. Epoxy cements set up rapidly and provide an excellent bond to clean, dry metallic surfaces. Therefore, epoxy cements are never to be applied to dirty, wet surfaces. In addition, epoxy cements do not normally work well on plastic or flexible containers. Often, epoxy cements are applied over other temporary patches such as a plug or drift to secure them in place.

Plug-N-Dike® Putty

Plug-N-Dike® putty is a very thick, gummy substance which can be effectively applied to active leaks from containers of hazardous liquid chemicals. It is recommended for use on petroleum hydrocarbons and solvents.

Drifts and Wedges

Frequently, a leak from a container can be effectively stopped by using a drift or wedge manually hammered into the hole of the split in a container. Drifts and wedges can be wood, rubber, or even metal, plastic coated or dipped to provide some corrosion protection for acids. Normally, wood alone works well for many petroleum hydrocarbons and solvents. Some drifts and wedges are manufactured using sparkproof materials such as AMPCO® beryllium/copper alloys.

Lead Wool

Lead wool is lead in the form of familiar steel wool. It can sometimes be peened into small hairline cracks using sparkproof ballpeen hammers. Obviously, leaks under any sort of pressure and materials that react with lead are not candidates for this option.

Overpacking a Patched Container

A patched container cannot be guaranteed not to leak again. It is a good policy to overpack such a container into a suitable recovery or salvage drum. The recovery drum selected must be constructed of a material compatible with the hazardous chemical involved. Overpack drums are available in steel, polyethylene, and epoxy/fiberglass materials. Care must be taken not to overpack drums upside down because it will be necessary later for someone to access the bung openings on the inner drum. Flammable liquid drums must never be overpacked when flammable vapors are present.

Pipe Clamps

Compression pipe clamps work well in stemming the flow from pipeline cracks or holes. Pipe clamps are often constructed of stainless steel bands lined with a softer rubber or neoprene inner liner. To prevent degradation of the rubber liner, an insert of Teflon® or other material is often placed over the liner to afford better chemical resistance.

Pipe Test Plugs

Emergency response personnel can use pipe test plugs to close off leaking, open-ended pipes of various diameters. These plugs are often inserted with the plug's center pipe remaining in the open position. This condition allows fluid flow and relieves pressure during installation. After installation, the valve on the pipe is slowly closed. Pipe test plugs are often constructed of aluminum or steel bodies with the expandable plug made of natural or neoprene rubber. For materials which react with the rubber, a flexible, chemically resistant sheathing can be applied over the end of the test plug before installation.

EMERGENCY RESPONSE PROCEDURES FOR FLAMMABLE CHEMICALS

Flammable chemicals may be gases, liquids, or solids. Recommended emergency response actions for flammable chemicals in each physical state are discussed below:

Flammable Gases

Flammable gases are often ignited immediately following a breach of their container. If the gases are not immediately ignited after sudden release, they can and often do flow until an ignition source is reached. When this happens, the entire cloud can catch fire almost at once. A gas leak which has ignited should not be extinguished unless the leak can be stopped. Attempts to shut off the supply should be made only under streams of water to keep the metal container cool. The gas should be allowed to burn itself out if the valves cannot be closed.

Flammable gases stored in metal cylinders are in liquefied form. When they vaporize under normal temperature and atmospheric conditions, they produce anywhere from 200 to 800 volumes of gas for each volume of liquid. The flammability range, which varies for each gas, is an important factor in emergency response. Some of these gases are flammable in mixtures as lean as 2%. In other words, gas diluted by 50 volumes of air to one volume of gas could be flammable. Explosimeters or other detectors must be used to check for dangerous concentrations. Continued monitoring on the lower flammability limit (LFL) is required until the release is stopped and the gas is known to have dispersed.

Flammable Liquids

As discussed earlier, the most important parameter to be considered in flammable or combustible liquids is their flash point. The possibility of ignition is greatest for liquids having low flash points. The lower the flash point, the greater the probability that either the temperature of the liquid or the temperature of the surrounding air will be higher than the flash point of the liquid. The higher the temperature of these, the greater the amount of vapor that could be formed, and the greater the hazard. Many chemical emergency response methods are available for flammable liquid spills and fires. In addition to some of the methods discussed, the following applications are available:

Dry Chemical. For small flammable liquid fires, portable dry chemical extinguishers are very effective. Directed at the heat of the fire, a dry chemical extinguishing agent interferes with the complex chemical chain reaction of the burning liquid and extinguishes the fire. Large fires can be substantially controlled by dry chemicals until other extinguishing agents are available.

Water. For large fires the most effective use of water is for exposure cooling and not for extinguishment. Some of the shortcomings of water are obvious, such as being heavier than most flammable liquids. Overuse or total reliance on water can promote the spreading of the burning liquid and cause frothing in heavy oils.

Water Fog. As water changes to steam it absorbs more heat than any other extinguishing agent. This heat of vaporization is known as latent heat. The fine droplets of water fog provide a large surface area for rapid heat absorption and have been used to reduce heat intensity. Water fog is effective as an extinguishing agent for flammable liquids with flash points above 100°F (38°C). The higher the flash point, the more effective water fog is in extinguishment.

Foams. Foams form a film on the surface of hydrocarbon fuel or a gummy membrane on a polar solvent. Other types of foams exist in addition to the HAZMAT® and Universal® foams already discussed. At one time protein foam was considered the most effective agent for flammable liquid fires. Today, protein foam has been replaced by aqueous film forming foam (AFFF). AFFF is used for large fires, and is best for flammable liquids which have a rather low flash point such as gasoline, –45°F (–43°C).

Flammable Solids

Flammable solids are likely to cause fires through friction or retained heat from manufacturing or processing, or can be ignited readily. Ignited flammable solids burn so vigorously and persistently that they may create a serious hazard. Flammable solids include both spontaneous combustible and water-reactive materials.

When charcoal is on fire, water should not be used immediately to extinguish the fire. If it is practicable, locate and remove the material on fire. The reason for this procedure is that wet charcoal may ignite spontaneously unless copious amounts of water are used. If this is not practical, copious amounts of water may be used to extinguish the visible fire. All charcoal is then removed and, if practical, the wet is separated from the dry material. The wet charcoal must be destroyed. The dry charcoal should be stored under cover in a dry place, and held under observation in order to avoid reignition.

Phosphorus will ignite with air at or above 86°F (25°C). Phosphorus becomes explosive when mixed with oxidizing materials. If a container of phosphorus is opened by accident, fire will occur. Therefore, any opened container of phosphorus should be filled with water. Phosphorus fire should be fought with dirt or sand, preferably wet. The residue from burned phosphorus is phosphorus pentoxide, a corrosive material that will dissolve in water to form phosphoric acid, another corrosive material.

Alkali metals such as lithium, sodium, and potassium are water-reactive. Do not use water on these metals or other water-reactive materials. Other fire extinguishing agents that must not be used include: carbon dioxide, carbon tetrachloride, foam, or any other liquid extinguishing agents such as freons, because they will react violently. Some of the agents that can be used include: dry soda ash, dry earth, dry salt lime, dry powdered limestone, or any dry inert material.

EMERGENCY RESPONSE PROCEDURES FOR OXIDIZERS

Oxidizers contain a considerable amount of oxygen in their molecular make-up and do not retain their oxygen atoms well. At times, it appears that only a fine line separates flammable solids and oxidizers. The heat of a fire can start an oxidizer releasing its oxygen atoms. A sudden release of oxygen enriches the atmospheric content of oxygen to above 21%. If ordinary combustible materials burn well in air, they will burn furiously in oxygen of higher concentration. Therefore, oxidizers increase the combustion rate of materials. Another danger associated with organic oxidizers is that they themselves are also flammable. Organic oxidizers which release oxygen can suddenly burn explosively.

Several oxidizers can cause fire when in direct contact with combustible materials. Examples of such oxidizers are nitric acid, perchloric acid, and concentrated hydrogen peroxide. Solid oxidizers that are mixed with finely divided combustible material may burn with almost explosive violence. In accidents involving chlorates, care must be taken to prevent ignition by friction or by contact with acid. Chlorates mixed with organic matter, or even with dust, form very flammable mixtures. Chlorates in contact with sulfuric acid can cause fire or explosion.

Fires caused by oxidizers are best extinguished by flooding with water. Copious amounts of water must be used to extinguish the fire and to dilute and wash away the liquid. Contact of the water with acids may cause splattering, and slight explosions can occur. Thus, a stream of water should be applied to the fire from a safe distance. Fumes caused by fires of nitric acid or mixed acids are poisonous and irritating. Workers must not be exposed to such fumes without appropriate personal protective apparel and breathing equipment.

Ammonium nitrate, sodium nitrate, other nitrates and ammonium nitrate fertilizer will not burn alone but when intimately mixed with organic matter will burn explosively if ignited. The burning heat melts the nitrate which may then ignite any combustible material it comes into contact with. Melted nitrate holds a great deal of heat. If water is thrown on melted nitrate, the sudden generation of steam will cause the melted nitrate to scatter and a fresh fire may result. Whenever practicable, nitrate fires must be smothered with earth, sand, or some other inert material. Water may be used if the fire is small. The use of water on a large nitrate fire is seldom effective.

EMERGENCY RESPONSE PROCEDURES FOR ORGANIC PEROXIDES

Organic peroxides may be either liquids or solids. Organic peroxides support the burning of combustible materials. The special property generally present in organic peroxides may cause their explosion if heated beyond their transportation temperatures. Most organic peroxides are shipped at ambient temperature, but a few must be shipped under refrigeration. If the refrigeration fails, the product may decompose and give off sufficient heat to start a fire.

Liquid organic peroxides take some effort to ignite. Once ignited, liquid organic peroxides burn with increasing rapidity as the fire progresses. If spilled on combustible materials, spontaneous ignition of the material may occur. Contamination of organic peroxides with a wide variety of chemical substances can cause a violent chemical reaction.

Solid organic peroxides are more readily ignited and burn with increasing rapidity as the fire progresses. If mixed with finely divided combustible materials, explosive mixtures may be formed. In addition, solid organic peroxides are prone to ignition by friction. Workers must not be allowed to walk or wander through any spilled organic peroxide. Some of these materials can ignite on contact with organic substances, such as oils on the sole of the shoe. Solid organic peroxides may also ignite due to the friction caused by walking on them.

Picric acid slowly decomposes with age and can form explosive peroxide crystals under the lids of jars. The simple friction caused by unscrewing a lid may cause an explosion. Even disturbing a container may cause an explosion because of the friction caused by the molecules rubbing against each other.

Attacking an organic peroxide fire takes a lot of planning and study by the first responders and the fire department. It is true that some organic peroxide fires can be extinguished, but only when the first responders and the fire department have the resources, training, and technical assistance from experts.

EMERGENCY RESPONSE PROCEDURES FOR TOXIC CHEMICALS

According to the Department of Transportation (DOT) classification, toxic chemicals include Class A Poisons, Class B Poisons, and "irritating" materials. Class A Poisons are gases, and they have been discussed. Class B Poisons are liquids and solids, including pastes and semisolids. The vapors of some Class B Poisons can be dangerous or offensive but to a lesser degree than the gases, vapors, or liquids, requiring "Poison Gas" label or placard. Some Class B Poisons are also flammable and an expert should be consulted by the chemical emergency responder if at all possible.

All workers must be evacuated from the area known or suspected to be contaminated by any class of poisons or their vapors. Whenever possible, monitoring equipment must be used to aid in determining and evaluating the contaminated area. Proper personal protective apparel and breathing equipment must be worn by all emergency responders. Chemical emergency responders must be properly decontaminated after the incident. All contaminated equipment must be thoroughly cleaned and decontaminated before being put back into service.

Fire may be fought using conventional means but all personnel in the area must have appropriate personal protective apparel and breathing equipment. Since the runoff may be contaminated, precautions must be taken to contain and reduce the area affected by the runoff. The methods usually available, such as digging holes or building earthen dikes, are often effective. However, any chemical emergency response action plan must be thoroughly evaluated prior to an incident. Evaluation reduces not only the short-term problem but the potential post-accident cleanup costs and long-term environmental degradation as well.

If leakage occurs in a container holding toxic chemicals, no person may be allowed to enter the contaminated area without proper personal protection and breathing equipment. If trained and qualified persons are available, attempts to plug the leak to reduce the amount of toxic chemicals being released into the environment should be made. Every precaution must be taken to keep any toxic chemical from contacting the skin.

If the emergency involves chlorine, the Chlorine Institute has developed three different kits that are specifically designed to patch leaks from three different sizes of liquefied chlorine gas containers. The "A" kit is specifically designed to handle and control leaks in chlorine cylinders. The "B" kit is for ton containers, and the "C" kit is for tank cars or cargo tanks. These kits, however, are not designed for and cannot be used on compressed gases stored at high pressure.

EMERGENCY RESPONSE PROCEDURES FOR CORROSIVE CHEMICALS

Corrosive chemicals are liquids or solids which cause visible destruction or irreversible alterations in human tissue. In terms of tonnage, corrosive chemicals account for more tonnage in production than do flammable compressed gases. DOT statistics show that one third of all hazardous chemical accidents involve corrosive chemicals. Hazardous materials incidents involving corrosive chemicals rank second only to flammable liquid accidents in the U.S.

Emergency responders arriving at a corrosive chemical incident must maintain complete personal protection at all times. Vapors and fumes from ammonia, chlorine, hydrochloric acid, and many others can rapidly penetrate ordinary protective clothing and can cause irritation to any part of a damp body. Emergency responders must therefore avoid entering any vapor cloud of corrosive chemicals without proper personal protective apparel and breathing equipment.

Leaking acids and alkalis should be diked and controlled and should not be allowed to come in contact with one another. In addition, corrosive chemicals must not be allowed to come in contact with organic chemicals. For example, leaking concentrated sulfuric acid or oleum can be explosive on contact with diesel fuel. Under no circumstances should a chemical emergency responder practice "solution by dilution" or "neutralization" unless a specialist is on the scene. The neutralization concept is a good one but its application can get out of hand.

Most corrosive chemicals are inorganic and thus are nonflammable. Organic corrosive chemicals such a glacial acetic acid, acetic anhydride, and others, can and will burn. If the hazardous chemical incident involves inorganic corrosive chemicals, exposed organic chemicals could be oxidized to the point of ignition. If the inorganic chemical is involved in fire, the smoke can be heavily saturated with corrosive gases. The use of water may cause unexpected reactions and spread the hazard. If an organic corrosive chemical is on fire, let it consume itself. The primary objective in such a situation is to stop the flow of material, contain it, and isolate the area.

15 DECONTAMINATION OF EMERGENCY RESPONDERS

Decontamination is the process of removing or neutralizing contaminants that have accumulated on personal protective apparel and equipment. Proper decontamination is essential to ensure the health and safety of personnel responding to hazardous chemical incidents. Decontamination protects workers from those chemicals that may contaminate and eventually permeate the protective clothing, respiratory equipment, tools, and other equipment used on site. It protects all site personnel by minimizing the transfer of hazardous chemicals from contaminated areas into clean areas. In addition, it protects the community by preventing uncontrolled transportation of contaminants from the site of a hazardous chemical incident. The extent of its success depends on how well the chemical emergency response commander can control on-scene personnel and operation, how well the chemical emergency response personnel are trained, and the perceived seriousness of the necessary decontamination.

MINIMIZING CONTAMINATION

Chemical contaminants can be found in any physical state: solid, liquid, or gas/vapor. They can be located either on the surface of personal protective apparel or equipment, or permeated into the material. Surface contaminants are easily detected and removed. However, if contaminants which have permeated a material are not removed, they may continue to migrate through the material and diffuse or surface on the inner side of the material, causing unexpected exposure of personnel to the hazardous chemical.

The key to decontamination is to *avoid* contamination. The Standard Operating Procedures at any hazardous chemical incident are to train emergency responders to minimize contact with any chemical and reduce the potential for extensive contamination. The following points must be included in the training of emergency responders:

(1) Minimize or avoid direct contact with all hazardous chemicals.
(2) Work from the perimeter of a spill and avoid walking directly through the spill.
(3) Use remote handling and sampling techniques.
(4) Inspect carefully all personal protective equipment for integrity prior to use.
(5) Use disposable outer garments easily removed during decontamination.
(6) Use disposable equipment and tools.
(7) Protect air monitoring instruments in plastic bags but provide openings in the bags for sampling ports and sensors.
(8) Use large plastic bags to encase contaminated material and equipment, or other sources of contamination.

The chemical emergency responders must also be well trained in personal protection. For example, proper procedure for donning protective apparel and equipment will minimize the potential for contamination. As a rule, all fasteners must be used, zippers fully closed, all buttons used and all snaps closed. Gloves and boots must be tucked under the sleeves and legs of outer clothing, and hoods must be worm outside the collar. If necessary, all junctures must be taped to prevent contaminants from running inside the gloves, boots, and jackets.

DECONTAMINATION PLAN

A decontamination plan must be developed and set up before any personnel or equipment may enter any area where the potential for exposure to hazardous chemicals exists. The plan must define, at a minimum, the following elements:

(1) The number and layout of decontamination stations within the contamination reduction zone.
(2) Equipment needs for decontamination.
(3) Decontamination methods for personal protective apparel, equipment, and tools.
(4) Procedures for removal of personal protective apparel and equipment.
(5) Disposal methods for contaminated protective apparel, equipment, and tools.
(6) The flow route of personnel and equipment into and out of the contaminated areas.
(7) Procedures to prevent contamination of clean areas.

The decontamination plan is to be revised whenever changes occur in the type of personal protective apparel or equipment, the anticipated hazardous chemical incident, or any changes in technology or decontamination method.

DECONTAMINATION METHODS

All personnel, clothing, equipment, tools, and samples leaving the contaminated area must be decontaminated to remove any hazardous chemical which may have adhered to them. Decontamination methods can be either physical removal or chemical inactivation, or a combination of both methods.

Physical Removal

Gross contamination, in many cases, can be removed by physical means such as scraping, brushing, and wiping. Loose contaminants can be removed by the application of water. The application of water may reduce concentration but not the chemical composition. Needless to say, water reactivity must be considered before employing this technique. Removal of electrostatically attached materials can be enhanced by coating the clothing or equipment with anti-static solutions. Anti-static solutions are available commercially as wash additives or anti-static sprays. Some contaminants adhere by forces other than electrostatic attraction. Removal of such adhesive contaminants can be enhanced through certain methods such as solidification, freezing, melting, or by the use of sorbents. Volatile liquid contaminants can be removed from protective clothing or equipment by evaporation followed by a water rinse. Evaporation of volatile liquids can also be enhanced by using steam jets.

Chemical Inactivation

Chemical inactivation is the altering of the chemical structure of the hazardous materials, rendering them inactive. Commonly used agents include sodium hypochlorite (household bleach), sodium hydroxide (household drain cleaner), sodium carbonate (washing soda), calcium oxide (hydrated lime), or other household chemicals and detergents. Technical advice for chemical inactivation should be obtained from experts to ensure that the inactivation chemical used is not reactive with the contaminant.

Chemical inactivation reaction can pose health and safety hazards under certain circumstances. In pure form or at high concentration, decontamination solutions or chemicals may pose a direct health hazard to workers. In general, decontamination solutions and chemicals must be compatible with the clothing or equipment undergoing decontamination and with the removal of hazardous chemicals. Any decontamination solution or chemical that permeates, degrades, damages, or otherwise impairs the safe functioning of the protective clothing or equipment, is incompatible with the material or equipment, and should not be used.

TESTING FOR EFFECTIVENESS

Decontamination methods vary in effectiveness for removing different hazardous materials. The effectiveness of any decontamination method must be assessed at the beginning of a program and periodically throughout the lifetime of the program. If contaminants are not being removed or are penetrating protective clothing, the decontamination program must be revised.

There is no reliable test to determine the immediate effectiveness of decontamination. In some cases, the effectiveness can be estimated by visual observation. However, not all contaminants leave visible traces. Many contaminants can permeate clothing without leaving any visible signs. Certain contaminants, such as polycyclic aromatic hydrocarbons, fluoresce and can be visually detected when exposed to ultraviolet light. Use of ultraviolet light, however, can cause damage to the unprotected eyes. Therefore, a qualified health and safety professional must be consulted on the risks associated with ultraviolet light use at a hazardous materials incident operation.

For after-the-fact information on the effectiveness of decontamination, wipe sampling or cleaning solution analysis provides the tests. In the wipe sampling procedure, a piece of cloth, filter paper, or swab is wiped over the surface of the object to be tested and then analyzed in a laboratory. In the case of protective clothing, both the inner and outer surfaces must be tested. Cleaning solution analysis is another way to test the effectiveness of decontamination procedures. Elevated levels of contaminants in the final rinse solution may suggest that additional changing and rinsing are needed.

SAMPLE DECONTAMINATION PROCEDURE

Prior to initiating decontamination, the decontamination officer must select a suitable site to conduct decontamination of personnel and equipment. Site selection depends on many factors including: the nature of the hazardous chemical; the amount, location, and containment of the chemical; the potential for exposure; the potential for environmental damage; the proximity of incompatible chemicals; the movement of personnel and equipment among different zones; the available method for decontamination; the accessibility to the site; water supplies; and other factors. Unfortunately, often it is not possible to actually find an ideal site for decontamination. Real-life problems may force the movement of the decontamination site once it has been established and in operation.

At a hazardous chemical incident, decontamination facilities must be located in the contamination reduction zone, which is the area between the exclusion zone (contaminated area) and the support zone (clean area). Once a site has been designated, an isolation perimeter should be quickly established. Warning signs advising everyone involved of the hazard and identifying where contaminated personnel are to report for decontamination must be posted. If the decontamination site is located some distance from the incident, team transportation is an additional requirement. Vehicles will be necessary to carry the team of contaminated personnel and their personal protective apparel, SCBA, and equipment. Drivers of transportation vehicles must also wear the appropriate level of personal protection, including SCBA.

Decontamination procedures must provide an organized process by which levels of contamination are reduced. The decision to implement all or part of the procedures is based upon a field analysis of the hazards and risks involved. The typical hazardous chemical incident does not require an extensive or elaborate decontamination procedure. Variations are often necessary to solve the problems at hand. Creative thinking and good judgment concerning personnel safety and health must be exercised.

The decontamination process consists of a series of procedures performed in a specific sequence. For example, outer, more heavily contaminated items such as outer garments, outer boots and gloves, are to be decontaminated and removed first, followed by decontamination and removal of inner, less contaminated items. Each procedure must be performed at a separate station in order of decreasing contamination. Entry and exit points must be clearly marked.

Decontamination workers who initially come in contact with personnel and equipment leaving the exclusion zone will require more personal protection than those workers who are assigned to the last station of "clean" area. In most cases, decontamination personnel must wear the same level of personal protection as workers in the exclusion zone. In some cases, decontamination personnel may be sufficiently protected by wearing one level lower protection. Appropriate personal protective clothing and equipment for decontamination personnel should be selected by a qualified health and safety professional. All decontamination workers must themselves be decontaminated before entering the support zone. The extent of their decontamination should be determined by the types of contaminants contacted and the type of work performed.

The following 19-station decontamination procedure assumes the worst case of a scenario with the workers donning Level A protection (Figure 15-1). Again, it should be emphasized that this is only one way of addressing the issue of decontaminants.

Station 1: Entry Point and Tool Drop

An entry point should be established and marked clearly in order to guide contaminated personnel into the decontamination area. Any tool or equipment

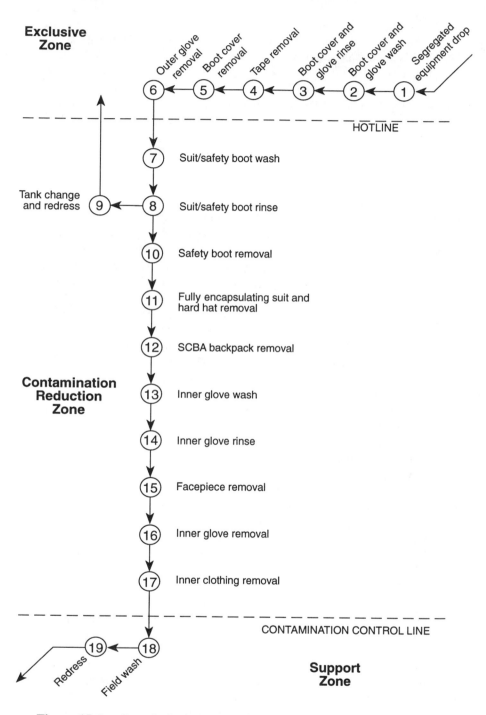

Figure 15-1.   Sample decontamination stations and layout.

which may be needed by other team members in the exclusion zone must be left at this point. This will isolate contaminated tools and equipment and make them available for replacement workers.

Station 2: Boot Cover and Glove Wash

Remove as much solid or liquid as possible by scrubbing outer boot covers and gloves with decontaminating solution or water and detergent. Runoff water must be diked or diverted to a safer area for treatment and disposal. At a minimum, keep runoff out of streams and ponds.

Station 3: Boot Cover and Glove Rinse

Rinse off decontaminating solution from Station 2 using copious amounts of water. Divert runoff water to a safer area for treatment and disposal.

Station 4: Tape Removal

Remove tape around boots and gloves and deposit in a container with plastic liner.

Station 5: Boot Cover Removal

Remove boot covers and deposit in a container with plastic liner.

Station 6: Outer Glove Removal

Remove outer gloves and deposit in a container with plastic liner.
The above six stations are set up on the edge of the exclusion zone. Beginning with Station 7, emergency responders now enter into the contamination reduction zone for further decontamination.

Station 7: Suit and Boot Wash

Wash encapsulating suit and boots using scrub brush and decontaminating solution or water and detergent. Repeat as many times as necessary. Divert runoff water to a safe area for treatment and disposal.

Station 8: Suit and Boot Rinse

Rinse off decontaminating solution from Station 7 using copious amounts of water. Divert runoff water to a safe area for treatment and disposal.

Station 9: Tank Change and Redress

If an SCBA air tank change is desired, it is the last step in the decontamination procedure. Air tank changed, new outer boot covers and gloves donned, joints taped, chemical emergency responders are redressed and can return to duty.

Station 10: Safety Boot Removal

If the chemical emergency responder does not have to return to duty, he can proceed from Station 8 to Station 10, remove safety boots and deposit them in a container with plastic liner.

Station 11: Fully Encapsulating Suit and Hard Hat Removal

A fully encapsulating suit is removed with assistance of a decontamination worker and laid out on a drop cloth or hung up. The hard hat is removed. Hot weather rest area may be set up within this station for personnel to cool off before returning to site.

Station 12: SCBA Backpack Removal

While still wearing facepiece, the chemical emergency responder removes backpack and places it on table. Contaminated SCBA must be isolated for complete decontamination at a later time. The hose is disconnected from regulator valve and the chemical emergency responder proceeds to the next station.

Station 13: Inner Glove Wash

Wash with decontaminating solution which will not harm the skin. Repeat as often as necessary. Divert runoff water to a safer area for treatment and disposal.

Station 14: Inner Glove Rinse

Rinse off decontaminating solution with water. Repeat as many times as necessary. Divert runoff water to a safe area for treatment and disposal.

Station 15: Facepiece Removal

Remove facepiece and deposit in container with plastic liner. Avoid touching face with fingers.

Station 16: Inner Glove Removal

Remove inner gloves and deposit in container with plastic liner.

Station 17: Complete Clothing Removal

Complete removal of personal clothing is required in many hazardous chemical incidents. This includes all undergarments and personal property such as rings, watches, wallets, and jewelry. All personal clothing will probably be disposed of, while metal items such as rings can be decontaminated.

Chemical emergency response personnel should consider removing personal items before entering contaminated area. Inner clothing must not be worn off-site since there is a possibility that small amounts of contaminants might have been transferred in removing the fully encapsulating suit.

Stations 7 through 17 are set up in the contamination reduction zone. Beyond Station 17, chemical emergency responders now enter into the support zone or clean area.

Station 18: Field Wash Area

Personal showering takes place at this station. Ample soap is to be applied to all areas of the body. Small brushes and sponges must be available for cleaning.
All cleaning items must be bagged for proper disposal and runoff must be controlled.

Station 19: Redress

Clean clothes are to be put on next. Disposable coveralls and hospital gowns and slippers are inexpensive and easy to use. These items should be prepackaged according to size and stored for immediate use.

Once personnel are thoroughly decontaminated, they proceed to a first aid station for medical evaluation. Vital signs must be recorded for each person leaving the decontamination area. Any open wounds or breaks in skin require immediate attention. Whenever all 19 stations in the decontamination procedure are activated, all decontaminated personnel leaving the hazard area must be transported to the appropriate hospital for further medical evaluation and monitoring.

When the decontamination process has been completed, the site must remain secure until the isolated clothing and equipment can be removed for proper decontamination, cleaning, or disposal. All runoff water and hazardous waste must also remain secure and ensure that a chain of custody is maintained. A debriefing must be held for those involved as soon as practical. Exposed chemical emergency responders and other personnel must be provided with as much data as possible about the delayed health effects of the hazards. Follow-up medical examination must be scheduled and exposure records maintained for future reference.

16   VICTIM HANDLING PROCEDURES

Following a hazardous chemical incident, early critical care life support is crucial to minimizing the potential morbidity and mortality. Providing effective care to victims, however, depends on the nature of the incident, the number of victims affected, the availability of medical care, the coordination of rescue efforts, the available modes of evacuation, and the availability of post-evacuation tertiary care. The goal in mass casualty scenarios is to minimize mortality and morbidity. Consequently, the most basic emergency response must include a method for assessing the incident, the extent of injury to victims, methods for determining which victims will receive treatment first, and what types of treatment will be given during the various stages of the incident. The basic need to handle hazardous chemical incident victims properly exists regardless of the environment in which the incident occurs, or whether small or large numbers of victims are involved.

STAGES OF EMERGENCY RESPONSE

Generally, chemical disasters have discrete stages, during each of which specific responses should occur. For the purpose of this chapter, five temporal stages are assumed:

(1) *Within Seconds*. Most likely, initial emergency treatment would be provided by bystanders and victims with minor injuries.
(2) *Within Minutes*. The magnitude of the disaster would have been determined, appropriate emergency response mobilized, and resuscitative victim care initiated.
(3) *Within Hours*. Definitive wound care and surgery to preserve lives would have been performed.

(4) *Within One Day.* Intensive care for the critically injured, combined care for less critically injured and treatment of early complication and sequelae would have been in place.
(5) *Within Two Days.* Potential community health issues associated with the disaster would have been identified and addressed.

ASSESSMENT OF VICTIMS

Initial response to any hazardous chemical incident is usually provided by bystanders or victims with minor injuries. The disaster site may be remote from essential health care resources, access to the site may be hampered, and prompt mobilization of rescue and medical personnel may be precluded. Consequently, bystanders or victims with minor injuries are realistically the only persons available to provide first aid in the critical moments following the disaster. It is probable the efforts of these individuals will be less than effective, since it is unlikely any of these individuals would have received prior training in first aid techniques. Ultimately, rescue and medical personnel will provide the initial medical care to the injured. Present-day hazardous materials response teams are comprised of technicians cross-trained to perform both the rescue and the medical management aspects of victim care.

Arrival of the first cohorts of emergency response and medical personnel must occur within minutes following the hazardous chemical incident. The role of the chemical emergency response personnel has been discussed. The responsibilities of the medical personnel are as follows:

(1) Determine the nature and severity of the injuries.
(2) Estimate the number of casualties.
(3) Provide initial treatment and resuscitations.
(4) Assess and triage disaster victims.
(5) Specify the resources needed to treat the injured initially at the site of the incident and request hospital resources for subsequent care.

In a mass casualty circumstance, the number of injured victims will exceed the number that may be cared for by the health systems operating under normal conditions. The matching of the severity of the injury with a health care facility able to treat the injury is of the utmost importance. Accurate triage decisions can maximize utilization of the resources in a manner that will increase survival of the victims when treatment resources are limited. There may be a high proportion of victims with multiple injuries to vital organs, suffering from internal and external hemorrhage, shock, decreased level of consciousness, or burns.

Field triage is based on information gathered by the triage officer and is usually limited to visual inspection of victims and a rapid assessment of the victim's vital signs. Ideally, victim assessment at the hazardous chemical incident site is conducted by emergency medical personnel, trained and certified to perform specific procedures. Realistically, the initial response to a hazardous chemical incident will almost always be lacking in adequately trained medical personnel, medical supplies, and modes of evacuation. Nevertheless, the goal of the chemical emergency response is to minimize morbidity and mortality, and effective care must be rendered as best as possible no matter how difficult the situation.

Individual victim assessment in the hazardous chemical incident setting is divided into two distinct phases: primary and secondary surveys. Primary survey deals with life-threatening injuries. In this phase, special attention is given to: a basic assessment and physical examination of the airway, breathing, and circulation (the ABC's); control of hemorrhage; and cervical and spine immobilization. This first step of rapid assessment includes the stabilization of the victim's cervical spine during the airway check and the immediate control of severe external bleeding identified during the circulation check. The primary survey is used to separate those individuals whose life-threatening injuries are treatable by swift intervention. It identifies those victims who would surely die if not treated immediately. Resuscitative maneuvers and immobilization of the cervical area should take place before moving the victim.

The secondary survey is performed after life-threatening injuries have been identified and further danger to the victim has been minimized. The purpose of the secondary survey is to identify any other less significant injuries which the victim may have suffered. The secondary survey is a systematic head-to-toe physical examination. It is particularly important for either the unconscious victim, or the victim with impaired hearing. Ideally, the secondary survey is completed at the staging area, well away from the actual hazardous chemical incident site. The information derived from this survey will serve to assign victims into the appropriate transport category.

TRIAGE

Triage refers to the sorting out and classification of disaster victims to determine priority of need and proper place of treatment to maximize the number of survivors. The technique of triage was first used by Napoleon's chief medical officer, who developed the principle of sorting patients for treatment based on medical need. The soldiers at risk of dying from their injuries received surgery first. Care of the less severely injured occurred after the gravely wounded were treated.

Modern triage of disaster victims gives priority to victims who will derive the most medical benefit from immediate treatment. Highest priority is given to victims who will live only if treated. Victims who will live without treatment and victims who will die despite treatment receive lower priority. In a disaster with mass casualties, the process of determining which victims should receive treatment first will often be the key to minimizing mortality. This task should be assigned to the senior officer not participating in victim treatment. Further application of a strategic triage system to identify those who need immediate treatment will optimize the use of available medical personnel and resources.

Information from the primary and secondary surveys can be used as the mechanism for the triage of disaster victims for immediate as well as definitive care along the following lines:

Immediate Treatment

Victims who have suffered life-threatening injuries and are in a critical yet potentially salvageable condition fall into this category. Frequently, these victims have suffered shock and severe blood loss, been unconscious, or have unresolved respiratory problems, severe chest or abdominal injuries, or major fractures. Also, three levels of burn severity are considered life-threatening and, therefore, warrant expeditious care and transport: burns associated with respiratory compromise, third-degree burns of more than 10% of the total body surface, and second-degree burns of more than 30% of the total body surface.

Second Priority Treatment

Victims who are considered to be urgently in need of treatment, but generally can be stabilized at the staging area with appropriate advanced life support interventions fall into this category. These victims should receive treatment within 2 hr. They include: those who have suffered back injuries, moderate loss of blood, or conscious head injuries. Also included are burned victims: those without respiratory compromise, those with third-degree burns of less than 10% of the total body surface, and those with second-degree burns of less than 30% of the total body surface.

Delayed Treatment

Victims who have suffered the least severe injuries, or at least have normal physiology following injury, fall into this category. Their treatment and transport is less urgent. Medical care can be delayed up to more than 2 hr. These injuries include minor fractures, burns, and soft tissue injuries.

Impending Death or Dead

Victims with no spontaneous respiratory or cardiac effort for more than 15 min and in whom cardiopulmonary resuscitation would be impossible due to the type of injury sustained are assigned to this category. This category is also reserved for victims who have suffered mortal wounds in which death appears imminent. Examples of such injuries include third- and fourth-degree burns of greater than 60% of the total body surface, coupled with other major injuries; or severe head or chest injuries.

INITIAL STABILIZATION

Successful treatment of the disaster victim depends on a pre-established systematic approach. The initial brief victim survey is directed toward correcting immediate, life-threatening problems of *a*irway, *b*reathing, *c*irculation *(the ABC's)*. To make a quick identification of problem areas on the conscious victim, place a finger on the victim's pulse and ask the victim a simple question. If the victim is alert with normal speech and pulse, proceed to the secondary victim survey, unless immediate eye or skin decontamination is needed. Detection of inadequate oxygenation or circulation demands immediate attention.

Priority 1: Evaluation of the Airway

Airway obstruction results from posterior displacement of the tongue, mucosal swelling, secretions, foreign bodies, and trauma. Airway patency must be assured by either chin lift or jaw thrust since hyperextension of the head and neck may aggravate any cervical spine injuries present. Secretions of the mouth should be cleared with a towel and any foreign bodies such as false teeth must be removed.

Priority 2: Evaluation of Breathing

Once an airway has been established, the respiratory system is checked by the "look, listen, and feel" method: *Look* for respiratory effort of watching for movement of the chest wall; *listen* for breathing by placing one's ear close enough to the victim's mouth and nose to hear air exchange; and concurrently *feel* for air movement against one's face. Respiratory rate is a sensitive measure of respiratory distress and should be counted for 15 sec and multiplied by 4 to give the rate per minute. The depth of respiration should also be noted.

Priority 3: Evaluation of Circulation

Circulation is checked by feeling for a carotid pulse located next to the voice box in adults and the brachial pulse located on the inner aspect of the upper arm in infants. Pulse rate and blood pressure should be obtained and recorded. While assessing for airway, breathing, and circulation, a rapid inspection of the total

Table 16-1   Glasgow Coma Scale

Component	Response	Score
Eye Opening	Spontaneous	4
	To verbal command	3
	To pain	2
	No response	1
Motor Response	Obeys command	6
	Localized pain	5
	Withdraw (pain)	4
	Flexion (pain)	3
	Extension (pain)	2
	No response	1
Verbal Response	Oriented	5
	Confused	4
	Inappropriate words	3
	Incomprehensible sounds	2
	No response	1
	Total	3 – 15

Note: The Glasgow Coma Scale is a simple way to evaluate and monitor the victim who is in coma from a head trauma. It has good value in predicting the outcome. Score each component by the best response. The higher the total score, the better the probability of survival.

body must be made to determine the presence of major hemorrhage. During the initial assessment, the chest wall should be evaluated so that obvious life-threatening injuries may be seen.

Priority 4: Assessment of Level of Consciousness

The victim's level of consciousness, and improvement or deterioration of same, is a valid and reliable survival predictor for trauma victims. The victim's level of consciousness and responses can be assessed by the Glasgow Coma Scale (Table 16-1). This information must be recorded with the other vital signs (pulse, blood pressure, respiration) and the time recorded on the victim's skin or tag. It is important to reassess the victim every 15 min or less if changes in condition become obvious.

Priority 5: Victim Decontamination

Cutaneous absorption is a common occurrence in exposure to hazardous chemicals. Cutaneous absorption depends on several factors such as lipid solubility, skin conditions, location on the body, nature of the chemical, and the presence of certain vehicles. Serious toxicity can result from cutaneous absorption. Chemical emergency responders must make every effort to reduce victim's absorption of hazardous chemicals. Chemical emergency responders must also take every precaution to protect themselves. In the case of victim eye contamination, immediately irrigate with copious amounts of water for at least 15 to 20 min with water or normal saline solution.

DECONTAMINATION OF VICTIMS

Decontamination of disaster victims is normally performed by immediately stripping and washing them down with copious amounts of water. Water alone, or water and detergent, are the most suitable agents for decontamination in almost all cases. On rare occasions, dry methods of cleaning using intrinsically safe vacuum cleaners for dry powder contamination, and scraping of tarry or viscous substances are employed. Emergency life-saving procedures take precedence over any other decontamination. It is extremely rare that a victim will be so heavily contaminated with such a highly toxic chemical that the rescuer's own life is put at risk.

The actual decontamination process basically consists of stripping and washing down the injured or contaminated casualties. All victims are to be afforded the most modesty possible for the circumstances, balancing this factor against the hazard involved. Contaminated clothing and personal effects should be kept together and labeled. Polyethylene bags are useful for clothes and personal effects. It is most important that any contamination is contained and confined as closely as possible to its site of origin. There may be occasions, however, when it is necessary for casualties to be removed from the scene to hospital for treatment while still contaminated. In such situations, every precaution must be taken to avoid contamination of the interior of the ambulance. This can be achieved by wrapping the victim in a polyethylene bag, leaving his head uncovered. Any ambulance that has been used to convey contaminated victims should not be used again until expert advice has been obtained about monitoring levels of contaminants and subsequent decontamination.

Some victims may take themselves to the hospital for treatment. It is important that any such victims are identified and recorded as being from the chemical incident. It is also necessary for hospital emergency departments to be prepared to receive injured and contaminated victims. Suitable rooms with adequate ventilation, extraction system, showers and portable baths are important. It is equally important that all other patients are kept away from contamination and further spread is kept to a minimum.

Once a victim has been declared dead by a physician, the body becomes the responsibility of the coroner. Depending on the nature of the hazardous chemical disaster, corpses can either be decontaminated at the scene by hosing them down or they can be enclosed in plastic at the scene and removed to the local mortuary.

All personnel handling corpses must be adequately informed of and protected against contamination. In major chemical disasters, temporary on-site mortuaries may need to be established.

EVACUATION OF VICTIMS

An evacuation plan for all potential hazardous chemical incidents must be prepared in advance and in coordination with the area disaster plan. Designated hospitals must be equipped and staffed to deal with contaminated casualties and those suffering from the acute effects of poisoning. A variety of suitable antidotes must be stored at these designated hospitals to deal with the anticipated cases of chemical poisoning. In the case of a major chemical disaster, hospitals must back each other up. Any victim of a chemical disaster must be evacuated only to designated hospitals unless life-threatening injuries preclude transport to such a designated hospital.

HEAT STRESS VICTIMS

Wearing both chemical-resistant protective clothing and self-contained breathing apparatus puts a worker at considerable risk of developing heat stress. The adverse effects to worker's health can range from transient heat fatigue to serious illness and even death. Heat stress is the burden or load of heat that must be dissipated if the body is to remain in thermal equilibrium. Heat stress is caused by a number of interacting factors, including environmental conditions, clothing, workload, and the individual characteristics of the worker. The physiological change resulting from heat stress is defined as heat strain.

Four environmental factors affect the amount of stress a worker faces in a hot environment: temperature, humidity, radiant heat, and air velocity. Essentially, the conditions worsen the hotter it is, the more humid it is, and the lower the air velocity. In addition, individuals vary in their susceptibility to heat stress. Personal characteristics such as age, obesity, alcohol and drug use, medical condition, physical fitness, chronic illness, and acclimatization to heat may predispose an individual to heat stress.

The body reacts to high temperatures by circulating blood to the skin, which increases skin temperature and allows the body to give off its excess heat through the skin. However, if the muscles are being used for physical labor, less blood is available to flow to the skin and release the heat. Sweating is another means the body uses to maintain a stable internal body temperature in the face of heat. However, sweating is effective only if a low humidity level exists to permit evaporation and if the fluid and salts lost are adequately replenished.

Increased risk of excessive heat stress is directly influenced by the amount and type of protective clothing and equipment worn. Personal protective clothing and equipment add weight and severely reduce the body's access to normal heat exchange mechanisms, i.e., convection, evaporation, and radiation. It is imperative that in the selection of protective clothing and equipment, each item's benefit must be carefully evaluated in relation to its potential for increasing the risk of heat stress. Once protective clothing and equipment are selected, the safe duration of work/rest periods should be determined based on the following factors:

(1) Type of protective ensemble.
(2) Ambient temperature and other environmental factors.
(3) Anticipated work rate.
(4) Individual worker characteristic and fitness.

If the body cannot dispose of excess heat, it will store it. When this happens, the body's core temperature rises and the heart rate increases. As the body continues to store heat, the individual begins to lose concentration and has difficulty focusing on a task, may become irritable, and often loses the desire to perform any task. This stress, usually in combination with other stresses such as workload, dehydration, and fatigue, may lead to heat disorders.

Heat Stress Injuries

Heat rash, also known as prickly heat, may occur in hot and humid environments where sweat is not easily removed from the surface of the skin by evaporation. *Heat syncope,* or fainting, may be a problem for the worker unacclimatized to a hot environment who simply stands still in the heat. Victims usually recover quickly after a brief period of lying down. Moving around, rather than standing still, will usually reduce the possibility of fainting. *Heat cramps* are painful spasms of the bone muscles. Heat cramps result when workers drink large quantities of water but fail to replenish the salt lost during sweating. Tired muscles are usually the ones most susceptible to cramps. Cramps may occur either during or after working hours. Cramps may be relieved by taking salted liquids by mouth or saline solutions intravenously for quicker relief.

Heat exhaustion develops as a result of loss of fluid through sweating when a worker has failed to drink enough fluids, take in enough salt, or both. The worker with heat exhaustion still sweats, but experiences extreme weakness or fatigue, giddiness, nausea, or headache. The skin is clammy and moist, the complexion pale or flushed and the body temperature normal or slightly higher. Treatment is usually simple: the victim rests in a cool place and drinks salted liquids. Severe cases involving victims who vomit or lose consciousness may require longer treatment under medical supervision.

Heat stroke is the most serious health problem for workers in hot environments. Heat stroke is caused by the failure of the body's internal mechanism for temperature regulation. Sweating stops and the body can no longer rid itself of excess heat. Signs include mental confusion, delirium, loss of consciousness, coma, with a body temperature of 106°F (41°C) or higher, and hot dry skin which may be red, mottled, or bluish. Victims of heat stroke will die unless treated promptly. While medical help should be called, the victim must be removed immediately to a cool area and his clothing soaked with cool water. The victim should also be fanned vigorously to increase cooling. Prompt medical attention can prevent permanent injury to the brain and other vital organs.

Because the incidence of heat stress depends on a variety of factors, all workers at hazardous chemical incidents, whether or not they wear any protective clothing and equipment, must be monitored for heat stress. For workers wearing permeable clothing, e.g., standard cotton or synthetic work clothes, follow the recommendations and the suggested work/rest schedules in the current edition of *Threshold Limit Values and Biological Indices* of the American Conference of Governmental Industrial Hygienists (ACGIH). If the actual clothing worn differs from the ACGIH ensemble in insulation value and/or wind and vapor permeability, change the monitoring requirements and the work/rest schedules accordingly. For workers wearing semipermeable or impermeable encapsulating ensemble, monitoring must be conducted when the temperature at the work site is above 70°F (21°C).

Heat Stress Monitoring

To monitor the worker, the following physiological monitoring should be conducted:

Heart Rate. Count the radial pulse rate as early as possible in the rest period. If the heart rate exceeds 110 beats per minute at the beginning of the rest period, shorten the next work cycle by one third and keep the rest period the same. If the heart rate continues to exceed 110 beats per minute at the beginning of the next rest period, shorten the following work cycle by one third.

Oral Temperature. Use a clinical thermometer or similar device to measure the oral temperature at the end of the work period but before drinking any water or liquid. If the oral temperature exceeds 99.6°F (37°C), do not allow the worker to wear a semipermeable or impermeable garment.

Body Water Loss. Measure worker's body weight on a scale accurate to ±0.25 lb at the beginning and end of each work day to see if enough liquids are being consumed to prevent dehydration. Body weights must be taken while the worker wears similar clothing or, ideally, very little clothing. The body water loss should not exceed 1.5% of total body weight in a work day.

Proper training and preventive measures will help avert serious illness and loss of work productivity. Preventing heat stress is particularly important because once a worker has suffered from heat exhaustion or heat stroke, that worker may be predisposed to additional heat injuries. To avoid heat stress, management must modify work/rest schedules according to physiological as well as environmental monitoring requirements; provide shaded areas or air-conditioned shelter to protect personnel during rest periods; maintain workers' body fluid levels; encourage workers to maintain an optimal level of physical fitness; and train workers in the recognition and treatment of heat stress.

ADMINISTRATION OF FIRST AID

The cardinal rule of administering first aid to hazardous chemical incident victims by emergency medical personnel is always to use personal protective and safety equipment relevant to the known hazard. Unconscious victims may require assisted ventilation and external chest compressions. Inhalation casualties must be removed to fresh air and allowed absolute rest where possible. Oxygen should be administered in high concentrations and at an appropriate flow rate for as long as is necessary. With smoke inhalation, there may also be burns to the respiratory tract which may not be immediately apparent until the secondary pulmonary edema develops. The appearance of such symptoms can be delayed for several hours. Victims of inhalation injuries must be evaluated by a physician.

Where there is chemical contamination, all affected clothes must be removed, together with shoes and wristwatches. The victim must be washed with copious amounts of water and soap. If the eyes are affected, they must be washed with normal saline. A simple arrangement is to use normal saline and a standard intravenous tubing set to irrigate the eyes continuously, with a tap to control flow. For hydrofluoric acid burns, a paste made from calcium gluconate must be used to cover the burn area. Organophosphorus compounds often produce symptoms that may require injections of atropine. Phosphorus gel ignites spontaneously in air. If a victim has phosphorus adherent to skin or clothes, the affected area needs to be submerged in water, or covered in very wet compresses or towels. In the hospital, the phosphorus can be removed under water by scraping with a spatula.

Inhalation First Aid

The following procedures are recommended for inhalation victims:

(1) Remove victim from further exposure.
(2) Check victim's breathing and pulse. Support ventilation, if necessary, using a self-inflating bag, not mouth-to-mouth, in cases of hydrogen cyanide or hydrogen sulfide exposure.
(3) Give external chest compressions, if victim does not have a pulse.
(4) If appropriate, position the unconscious victim who is breathing spontaneously on his side to prevent choking if vomiting occurs.

(5) If required, administer oxygen at 6–10 L by mask.
(6) Evacuate victim as soon as possible to hospital.

Skin Contamination First Aid

The following procedures are recommended for victims with skin contamination:

(1) Remove from victim all contaminated clothing, including shoes, watches, and jewelry.
(2) Drench affected area with copious amounts of water. If appropriate, also use soap or detergent.
(3) Check victim for chemical burns and signs of systemic poisoning.
(4) If victim is in shock, administer normal saline, or Ringer's lactate.
(5) Evacuate victim as soon as possible to hospital.

Eye Contamination First Aid

The following procedures are recommended for victims with eye contamination:

(1) Irrigate the eye with water or normal saline for at least 15 to 20 min.
(2) Evacuate victim as soon as possible to hospital.
(3) Refer ophthalmologist for eye examination.

Ingestion First Aid

The following procedures are recommended for ingestion victims:

(1) Initiate emesis with ipecac syrup except under the following conditions:

    (a) Ingestion of strong acids or bases; and
    (b) Victim is unconscious, has lost gag reflex, or is experiencing seizures.

(2) In the past, emesis after ingestion of petroleum hydrocarbons was thought to be contraindicated. However, emesis is now encouraged by most poison centers, depending on the amount ingested.
(3) Activated charcoal is to be administered following gastric evacuation. This bonds the poison, speeding its elimination from the gastrointestinal tract.

(4) If victim is unconscious, but has a pulse and is breathing normally, position the victim on his side.

(5) If victim is conscious, give a pint of water to drink immediately. If the chemical is corrosive, give a pint of milk unless the chemical contains phosphorus, chlorinated hydrocarbons, or degreasing solvents, in which case continue giving water.

(6) Save all vomits and send to the hospital should spontaneous vomiting occur.

(7) Evacuate victim as soon as possible to hospital.

17   LABORATORY WASTE MANAGEMENT

In 1976, Congress passed the Resource Conservation and Recovery Act (RCRA) which directed the U.S. Environmental Protection Agency (EPA) to develop and implement a program to protect human health and the environment from improper hazardous waste management practices. The program is designed to control the management of hazardous waste from its generation to its ultimate disposal, "from cradle to grave."

In implementing the RCRA regulations, EPA first focused on large companies, which generate the greatest portion of hazardous waste. Initially, business establishments generating less than 1000 kg (2200 lb) of hazardous waste in a calendar month, known as small quantity generators, were exempt from most of the hazardous waste management regulations.

In recent years, however, public attention has been focused on the potential for environmental and health problems that may result from mismanaging even small quantities of hazardous waste. In November 1984, the Hazardous and Solid Waste Amendments (HSWA) to RCRA were signed into law. With these amendments, Congress redefined the class of small quantity generators as those establishments which generate more than 100 kg but less than 1000 kg of hazardous waste in a calendar month. The final regulations for this redefined class of small quantity generators were issued in March 1986 and the requirements have been effective since September 1986. As a consequence of the regulations implementing the HSWA regulations, "conditionally exempt" small quantity generators were defined as those that produce less than 100 kg per month of hazardous waste and do not accumulate more than 1000 kg of hazardous waste at any one time. Conditionally exempt small quantity generators are exempt from all RCRA notification, reporting, and manifesting requirements. Conditionally exempt small quantity generators are, however, required to send their wastes to permitted or interim status treatment, storage, or disposal facilities, or to legitimate recycling or reclamation facilities.

179

The discussion that follows in this chapter assumes the status of a small quantity generator under federal regulations. It is important to note that many states have different generator categories and requirements.

DEFINITION OF HAZARDOUS WASTE

A waste is any solid, liquid, or contained gaseous material no longer used. Waste is either recycled, thrown away, or stored until a sufficient quantity exists to be treated or disposed of. All laboratory work with chemicals eventually produces wastes that can cause serious problems if not handled and disposed of properly. If the laboratory waste generated can either cause injury or death, or pollute land, air, or water, then such waste must be considered hazardous. Hazardous wastes are regulated by federal and state public health and environmental safety laws.

There are two ways a waste may be brought into the hazardous waste regulatory system: the waste is either specifically listed as hazardous, or it is identified as hazardous through characteristics. Congress gives EPA the authority to list wastes generically under RCRA. If EPA has reason to believe that a laboratory waste is hazardous and meets a generic listing, the waste is automatically regulated. The waste may be "delisted" if the organization can successfully demonstrate that the waste is nonhazardous according to the delisting criteria. EPA may list a waste as hazardous if the waste "typically and frequently" falls into one of three general categories:

(1) *Containing Hazardous Characteristics.* The four hazardous characteristics are: ignitability, corrosivity, reactivity, and TCLP (toxicity characteristic leaching procedure) toxicity. These characteristics and their tests will be discussed later.

(2) *Acutely Hazardous.* The waste has been shown to be fatal to humans in low doses. EPA uses toxicity information from the literature to decide which wastes are acutely hazardous.

(3) *Toxic.* EPA lists wastes as "toxic" if the wastes contain certain "hazardous constituents" and EPA judges that the waste could pose health or environmental problems according to certain criteria including chronic toxicity, migration potential, bioaccumulation, quantities of waste generated, and past environmental problems caused by mismanagement of the waste.

EPA-Listed RCRA Hazardous Wastes

There are four lists of hazardous wastes (Appendix E):

(1) *Nonspecific Source Wastes.* These are wastes which cannot be attributed to any one industry. These wastes are assigned EPA hazardous waste numbers of three digits preceded by "F".

(2) *Specific Source Wastes.* These are wastes from specific industries that EPA has determined to be hazardous. For example, there are specific source wastes for the following industries: chemical manufacturers, coking, iron and steel foundries, pesticide manufacturers, and others. These wastes are assigned EPA hazardous waste numbers of three digits preceded by "K".

(3) *Discarded Commercial Chemical Products — Acutely Hazardous.* The chemicals on this list are primarily acutely hazardous. Empty containers of these chemicals are hazardous wastes as well unless the container's liner has been either removed, or the container has been decontaminated. These discarded commercial chemical products are assigned EPA hazardous waste numbers of three digits preceded by "P".

(4) *Discarded Commercial Chemical Products — Hazardous.* The chemicals on this list are primarily toxic. These discarded commercial chemical products are assigned EPA hazardous waste numbers of three digits preceded by "U".

Hazardous Waste Characteristics

Even if a waste does not appear on one of the four EPA lists, it is considered hazardous if it has one or more of the following characteristics:

(1) *Ignitability.* The characteristic of ignitability can apply to any physical state: solid, liquid, or containerized gas. The test of ignitability is whether the waste has a flash point of less than 140°F (32°C). Excluded from this test are aqueous liquids containing less than 24% alcohol by volume, since these liquids will often have a low flash point but the water content keeps the liquid from supporting combustion. Wastes that exhibit the ignitability characteristics are assigned the EPA hazardous waste number D001.

(2) *Corrosivity.* The characteristic of corrosivity applies only to liquids. A waste is corrosive if its pH is less than 2 or greater than 12.5, or if it corrodes steel at a rate greater than 6.35 mm per year. Wastes that exhibit the corrosivity characteristic are assigned the EPA hazardous waste number D002.

(3) *Reactivity.* Reactivity is nebulous in character; no absolute tests exist to define it. This characteristic is intended to identify those wastes that have the potential to explode or release toxic gases

during the waste management process. The intent is twofold: to protect the people working with the waste, and to prevent fires and explosions. EPA has issued a guidance standard that a waste is RCRA reactive if OSHA limits for cyanide or sulfide have been exceeded in areas where the waste is generated, stored, or otherwise handled. It should be noted that the characteristic of RCRA reactivity is different from DOT reactivity. Wastes that exhibit the RCRA characteristic of reactivity are assigned the EPA hazardous waste number D003.

(4) *TCLP.* TCLP is an acronym for "toxicity characteristic leaching procedure," a laboratory procedure which produces a leachate from a waste. It is the standardized method adopted by the EPA to determine if wastes exhibit the toxicity characteristic. TCLP replaces the former EP Toxicity test, (EP stands for "extraction procedure"), and allows for the testing of organic constituents that were not included in the former EP Toxicity test. TCLP is a laboratory test used to simulate landfill conditions, and to determine whether a waste contains a regulated toxic constituent which would leach from the waste under landfill conditions, resulting in the potential contamination of groundwater. Wastes that exhibit the characteristic TCLP toxicity are assigned the EPA hazardous waste numbers listed in Table 17-1 if the associated maximum contaminant level is exceeded.

Mixed Wastes

All laboratory personnel must be aware of the very specific rules dealing with mixtures containing hazardous wastes:

(1) The entire mixture of listed hazardous waste with nonhazardous waste will remain regulated as that listed waste, regardless of the quantity of the listed waste present in the mixture.

(2) A mixture of characteristic hazardous waste with nonhazardous waste will be regulated as hazardous only if the mixture exhibits such a characteristic.

Spill Cleanup

Chemical emergency response and spill cleanup often produce contaminated materials that become wastes and need to be disposed of. These materials may be residue, contaminated soil or water, rinsings, absorbent, and other debris. The rules on mixed wastes will apply in determining whether a waste is hazardous.

Table 17-1   Criteria for TCLP

EPA Hazardous

Waste Number	Contaminant	MCL (ppm)
D004	Arsenic	5.0
D005	Barium	100.0
D006	Cadmium	1.0
D007	Chromium	5.0
D008	Lead	5.0
D009	Mercury	0.2
D010	Selenium	1.0
D011	Silver	5.0
D012	Endrin	0.02
D013	Lindane	0.4
D014	Methoxychlor	10.0
D015	Toxaphene	0.5
D016	2,4-D	10.0
D017	2,4,6-TP Silvex	1.0
D018	Benzene	0.5
D019	Carbon tetrachloride	0.5
D020	Chlordane	0.03
D021	Chlorobenzene	100.0
D022	Chloroform	6.0
D023	o-Cresol	200.0
D024	m-Cresol	200.0
D025	p-Cresol	200.0
D026	Cresol	200.0
D027	1,4-Dichlorobenzene	7.5
D028	1,2-Dichloroethane	0.5
D029	1,1-Dichloroethylene	0.7
D030	2,4-Dinitrotoluene	0.13
D031	Heptachlor (and epoxide)	0.003
D032	Hexachlorobenzene	3.0
D033	Hexachloro-1,3-butadiene	0.5
D034	Methoxychlor	10.0
D035	Methyl ethyl ketone	200.0
D036	Nitrobenzene	2.0
D037	Pentachlorophenol	100.0
D038	Pyridine	5.0
D039	Tetrachloroethylene	0.7
D040	Trichloroethylene	0.5
D041	2,4,5-Trichlorophenol	400.0
D042	2,4,6-Trichlorophenol	2.0
D043	Vinyl chloride	0.2

Empty Containers

A container is considered empty by RCRA definition if any one of the following three conditions is met:

(1)   All possible material has been removed from the container using common practices, e.g., pouring, pumping, etc., and no more than one inch of residue is contained on the bottom of the container.
(2)   For containers of 110 gal or less, remaining residue is no more than 3% by weight of the total capacity of the container.
(3)   For containers greater than 110 gal, remaining residue is no more than 0.3% by weight of the total capacity of the container.

An empty container is not regulated as RCRA hazardous waste with the exception that an empty container that has held an acute hazardous material is hazardous waste unless the liner is removed or the container has been triple-rinsed with a suitable solvent. Liners or rinsings from container decontamination are regulated as hazardous wastes.

CATEGORIES OF GENERATORS

In March 1986, the federal rules for hazardous waste management were modified to bring organizations that generate small amounts of hazardous waste into the regulatory system. Previously, small quantity generators that generated less than 1000 kg (2200 lb) of hazardous waste in a calendar month has been exempt from most hazardous waste regulations. There are three categories of hazardous waste generators:

Full-Compliance Generators

If an organization generates 1000 kg (2200 lb), or 300 gal or more, of hazardous waste, or more than 1 kg of acutely hazardous waste in any month, this organization is required to comply with *all* applicable EPA hazardous waste management rules.

Small Quantity Generators

If an organization generates more than 100 but less than 1000 kg (220 to 2200 lb), or 30 to 300 gal, of hazardous waste and no more than 1 kg of acutely hazardous waste in any given month, the organization is required to comply with the 1986 EPA rules for managing hazardous waste, including the accumulation, treatment, storage, and disposal requirements which will be described later.

Conditionally Exempt Small Quantity Generators

If an organization generates no more than 100 kg (220 lb) or 30 gal of hazardous waste and no more than 1 kg of acutely hazardous waste in any calendar month, federal hazardous waste laws require the organization to:

(1) Identify all hazardous waste it generates;
(2) Send waste to a hazardous waste facility, or a landfill or other facility approved by the state for industrial or municipal wastes; and
(3) Never accumulate more than 1000 kg of hazardous waste on the property. If hazardous waste is accumulated more than 1000 kg on the property, the organization will become subject to all the requirements of a small quantity generator.

Under the federal hazardous waste management system, an organization may be regulated under different rules at different times, depending on the amount of hazardous waste it generates in a given month. For example, a conditionally exempt small quantity generator will become a small quantity generator if the organization generates between 100 and 1000 kg of hazardous waste in any month. Once the 1000 kg/month threshold of hazardous waste generation is crossed, the organization will be subject to all applicable hazardous waste management regulations, as will all other hazardous wastes it generated in previous months.

ON-SITE WASTE STORAGE

Small quantity generators may store no more than 6000 kg of hazardous waste on site for up to 180 days, or for up to 270 days if the waste must be shipped to a treatment, storage, or disposal facility which is located over 200 mi away. Small quantity generators are allowed to store for as long as 180 or 270 days so that they will have time to accumulate enough hazardous waste to be shipped off-site. If a small quantity generator exceeds either the quantity or time limit, the organization will be considered a storage facility and it must obtain a storage permit and meet all of the RCRA storage requirements. The time limit for full-compliance generators is 90 days.

All hazardous waste stored on-site must be stored in containers suitable for the type of waste generated in order to protect human health and the environment, and reduce the likelihood of damages or injuries caused by leaks or accidental spills. The following guidelines are recommended for on-site hazardous waste storage in containers:

(1)   Each container must be clearly marked with the words "HAZARD-OUS WASTE", and with the date it begins to accumulate.

(2)   All hazardous waste storage containers must be in good condition and without any leaks.

(3)   Hazardous wastes are not to be stored in any container that may result in corrosion, leaks, rupture, or other failure.

(4)   Hazardous waste storage containers must be kept closed at all times except for filling or emptying.

(5)   Hazardous waste storage containers must be inspected periodically for corrosion, leaks, rupture, or other failure.

(6)   Containers of ignitable or reactive wastes must be placed as far as possible from the facility property line to create a buffer zone to protect the public.

(7)   Incompatible wastes must never be stored in the same container to avoid potential chemical reaction that may cause an explosion, fires, leaks, release, or other danger (Appendix G).

(8)   Stored wastes must be shipped off-site or treated on-site within 180 (or 270) days.

HAZARDOUS WASTE SHIPMENT PREPARATION

Most laboratories will dispose of hazardous laboratory wastes off-site in an EPA-permitted hazardous waste landfill. Since the ultimate responsibility for damages caused by a waste is borne by the generator, it is very important that a secure, permitted treatment/disposal facility be used. Most disposal facilities will accept a variety of wastes including solids, liquids, and "lab packs". Liquids must be containerized and solidified prior to landfilling. Hazardous wastes must be put in containers acceptable for transportation prior to shipment. All hazardous waste containers must be clearly and properly labeled.

A "lab pack" is a DOT specified 17H steel drum of 55-gal capacity with removable head (lid) that is filled with waste materials in purchased or stored packages (Figure 17-1). Regulations require inert packing material to be used to surround each inside package in order to separate and maintain the integrity of the small bottles and vials containing the wastes. Packing material must be sufficient to absorb any liquid released from broken or leaking internal packages. The maximum volume of waste contained in the "lab pack" is about 15 gal.

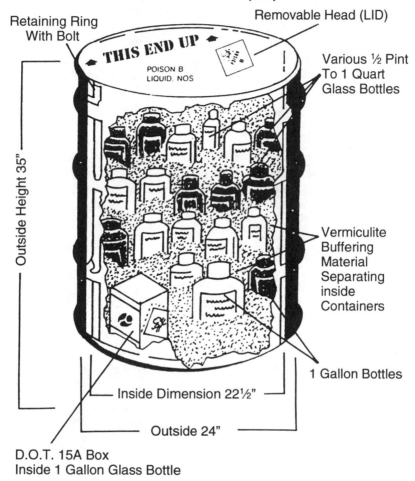

**Drum—Typical D.O.T. 17H Steel
With Removable Head (LID)**

Retaining Ring
With Bolt

Removable Head (LID)

THIS END UP

POISON B
LIQUID. NOS

Various ½ Pint
To 1 Quart
Glass Bottles

Outside Height 35"

Vermiculite
Buffering
Material
Separating
inside
Containers

1 Gallon Bottles

Inside Dimension 22½"

Outside 24"

D.O.T. 15A Box
Inside 1 Gallon Glass Bottle

Figure 17-1   A "Lab Pack".

THE UNIFORM HAZARDOUS WASTE MANIFEST

A uniform hazardous waste manifest is a multicopy shipping document which a generator must fill out and use to accompany its hazardous waste shipments. The manifest form is designed for tracking of shipments of hazardous waste from point of origin to final destination, the so-called "cradle-to-grave" system. The hazardous waste generator, the transporter, and the designated treatment/storage/disposal facility must each sign this document and keep a copy. The treatment/storage/disposal facility operator must also send a copy back to the generator, to assure the generator that his shipment arrived at its destination. It is required by law that the generator must keep this copy of manifest, signed by both the transporter and the facility, on file for 3 years.

If the generator does not receive a signed copy from the facility within 30 days, it must be reported to the state or federal EPA. Even though the hazardous waste may have been picked up by the transporter and is no longer in the generator's possession, his liability does not end there. The generator is still liable under Superfund for any mismanagement of the hazardous waste by either the transporter or the treatment/storage/disposal facility. Therefore, the importance of careful selection of a transporter and a treatment/storage/disposal facility can never be overemphasized.

WASTE REDUCTION

Under the Hazardous and Solid Waste Amendments (HSWA) of 1984, all generators must certify on the uniform hazardous waste manifest that they have in place a program to reduce the volume or quantity and toxicity of the waste they generate to a degree determined by the generator to be economically practicable. Generators must also certify that their current method of management is the most practicable method available to minimize present and future threat to human health and the environment.

18   EMPLOYEE TRAINING

An effective occupational safety and health program is based on a competent job performance. Workers who are properly trained to do their jobs will do them safely. Training is only one way to influence the behavior of workers. An employer who invests a great deal to create safe working conditions will encourage employees to perform safely. Safe performance is also promoted by developing safe work procedures, by teaching the procedures effectively, by insisting that the rules are followed, and by teaching the facts regarding accident causes and preventive measures. A well-planned training program not only will train employees, but also provide positive influences which further complement the effect of the training.

TRAINING PROGRAM DEVELOPMENT

When a training program is to be developed, the following points must be considered:

Program Needs and Objectives

A training program is needed when: new workers are hired; workers are transferred to a new department; new equipment or processes are being introduced; operating procedures are being revised; new information is to be disseminated; or worker performance needs improvement.

Employee training programs should be based on clearly defined objectives that determine the scope of the training and guide the selection and preparation of the training materials. Training program objectives should state what the trainee is expected to know or do upon completion of the program. Any employee training program must obviously have management approval, as well as full acceptance by the supervisors. Yet, in many organizations top management must also be targeted for training. In order for any employee training program to succeed, all three groups — top management, middle management, and the worker — must give both approval and acceptance.

Training Outline and Material

After defining the training program objectives, the next step is to develop an outline of what is to be covered in the program. Sometimes, a training outline may meet the program objectives by using existing texts and course materials, or by combining parts of several texts or courses already in use. Frequently, however, a completely new program must be designed. Either way, the training outline must be developed in detail. Major topics and subtopics should be logically arranged and the time allocated to each topic should be in proportion to the importance of each.

Today, there are commercially available a variety of audiovisual media that are produced for employee training. Such "canned" presentations may be in the form of a slide/tape program, film, videotape, or some other format. The canned presentation is one way of ensuring consistency in training and that the major points are covered. In addition, the presentation can be shown to large or small groups. Their major disadvantage is the absence of flexibility unless supplemented with the organization's own in-house materials to personalize the presentation.

Training Methods

The methods of training depend on a number of factors. Among them are: the trainer; the trainee; the training content; the size of the group to be trained; as well as other related factors. The type of training usually falls into the following categories: (1) individual on-the-job training, (2) small-group instruction, and (3) large-group instruction.

The best time to offer the training is during regular working hours. New employees should be trained prior to assuming their work assignments or within 1 month of their employment. Employees who change job responsibilities within the organization should be trained within 1 month of assuming their new responsibilities. Training should also be given on an annual basis as a refresher.

THE TRAINER

The person who performs the training must be a competent information source and possess the ability to teach. Chemistry instructors, industrial hygienists, safety professionals, and well-informed supervisors have different educational backgrounds but may be able to perform the function of communicating safety information that is at the heart of the training. The following are important characteristics to look for in a trainer:

(1) *Enthusiasm.* The trainer must have the desire to share information and help the worker in the learning process.

(2) *Communication Skills.* The trainer must have the ability to organize and present information in a clear manner. The trainer must make the information understandable to the worker.

(3) *Content Competency.* The trainer must be competent in the subject matter being taught.

It is important to remember that employees are students and as such the trainer must view their training from a teacher-student prospective. Remember also that this type of training is most effective when conducted in nontechnical language. This is especially important for those employees who are not scientifically oriented and are not familiar with some of the technical language or jargon.

TRAINING FOR SUPERVISORS

The immediate job of preventing laboratory accidents and controlling safety and health hazards falls upon the supervisors. Safety and production are part of the supervisory function. Simply getting supervisors to agree in theory that safety is one of their duties is not enough. Supervisors must come to understand the many ways in which they can prevent accidents, and become interested in improving safety and health performance.

Some supervisors proclaim that the knowledge and philosophy of accident prevention is just common sense. These individuals fail to realize that safety knowledge is a specialized body of information accumulated over a period of many years. It is the job of industrial hygienists and safety professionals to help supervisors gain whatever information is available to make their safety efforts more productive.

Success in supervision requires skill as much as it requires knowledge and understanding. Basic safety training is aimed at making all personnel aware of the need for safety programs, increasing safety knowledge, and promoting proper attitudes regarding safe behavior. Many job accidents are the direct result of unskillful handling of the important tasks of supervision, such as giving orders, assigning jobs, correcting workers for poor performance, and the like. In short, to do a good safety job, supervisors need training in supervisory skills. Among the most important of these skills are giving job instruction, work supervision, determining accident causes, and building positive safety attitudes among workers.

NEW EMPLOYEE TRAINING

Effective laboratory accident prevention begins on the first day of a worker's employment and continues for the employee's entire work career. Since first impressions are usually long-lasting ones, safety and health training programs for new employees are very important. Every organization must endeavor to impress new employees with the organization's concern for safety and health. Absolutely no doubt about these concerns is to exist. Possible doubts are erased by stressing the safety and health aspects of all laboratory operations during their training period.

The amounts of time and effort devoted to safety and health in new employee training sets the tone for the new employee's level of safety and health awareness. Today, it is commonplace for organizations to demonstrate concern for employee health and safety by making a positive first impression by devoting the employee's first day to the coverage of health and safety items.

The orientation session provides each employee the opportunity of exposure to the organization's safety policy. Yet the amount of information retained is usually limited. The natural nervousness of starting a new job and interest in other matters of immediate concern make it difficult for the employee to absorb safety instruction. Therefore, it is necessary to consider what safety information must be given first, and then the best way to present it. In order to have a good start in safety training, the following items must be included in the orientation program:

(1)   A management staff that is serious as well as sincere in an interest in accident prevention.

(2)   A health and safety program that is already in effect in the workplace, yet management will willingly expand the program.

(3)   The supervisor gives job instructions. No employee is expected to undertake a job before instruction on the method and the learning process is complete. The supervisor authorizes all work assignments.

(4)   If a job appears to be unsafe, the employee is to contact the supervisor. The supervisor will then consult with the worker.

(5)   Each employee is expected to report any unsafe conditions to the supervisor.

(6)   If an employee suffers an injury, no matter how minor, it must be immediately reported to the supervisor.

Preliminary safety and health instruction is most often given to new employees by the personnel department. It is advantageous for a management executive, a safety professional, or an industrial hygienist to give the safety instruction. More important than who gives the talk is how the talk is given. The talk should be prepared and presented with the utmost regard for the effect it will have on the new employees. Verbal instruction should be given earnestly and with an attitude of good will and friendly cooperation. The importance of the position of the supervisor in the safety program must be emphasized. The new employee must understand that the supervisor is responsible for job training, and such training will include instruction on safe work procedures. No gap or contradiction between the information given at orientation and that given later at the workplace is to exist. The supervisor must be aware of the content of the orientation.

A new employee reaching his own department will be given additional safety instruction by the supervisor. Instruction may cover some of the points made at the orientation, but is now more specifically applied to the work the employee will be doing. The supervisor usually repeats earlier safety instruction, but must instruct the new employee on the department's safety and health program. The supervisor will explain departmental safety regulations and will see to it that the new employee is provided with the appropriate personal protective equipment. In addition, the supervisor will make provisions for safe work procedure training and make certain that it will be followed.

Once the initial round of training has been completed, the new employee must then receive on-the-job training in his assigned work area. On-the-job training is widely used because the trainee can be producing while he is being trained. Whether the supervisor does the training himself or has someone else do it, the training must be carefully planned and organized. If the supervisor does not conduct on-the-job training himself, it is his responsibility to ensure the person delegated to do the training is capable of conducting training.

On-the-job training should not be a hit-or-miss situation. In too many cases, the new employee is simply told to follow another worker around and learn from that person. In situations of this type, the worker may be too busy to do any training; may not have been formally trained himself; or may not know how to train another person. He may even be reluctant to train another person who could potentially replace him in the future.

On-the-job training is the most flexible and direct of the three training methods. In on-the-job training, the new employee is developing and applying his skills in typical work situations under the personal guidance of a qualified person. Needless to say, the person to whom the trainee has been assigned should be one who knows the job and its safety and health issues thoroughly and is an experienced operator, with the patience and skill to train another person.

Other advantages of on-the-job training are: the trainer can identify specific performance deficiencies and when necessary take immediate and proper corrective action; the training results are readily apparent since the trainee is in a real work situation; and the trainee's progress can be judged continually.

After the employee has been oriented to the new job, it is the responsibility of the supervisor to maintain consistent safe work procedures, and, when necessary, discipline safety rule violations. If the supervisor observes workers taking shortcuts or otherwise departing from safe procedures, he must correct them at once. If the supervisor does not correct the employee on the spot, the unsafe procedure will soon become a habit.

TRAINING FOR TRANSFERRED EMPLOYEES

All employees transferred from one assignment to another, or from one department or shift to another, must receive initial safety and health training as well as periodic refresher training. The management must ensure that proper procedures are established and maintained for the training of transferred employees. Care must be taken to ensure that the necessary procedures are developed and training records are maintained.

KEEPING RECORD OF TRAINING ACTIVITIES

The importance of employee training is evidenced by the requirements of the Occupational Safety and Health Administration Safety and Health Training Requirements for General Industry (29 CFR 1910) and specifically in the Laboratory Standard (29 CFR 1910.1450). Accurate records must be maintained for all safety and health training activities. Training records must include the following information:

(1)   Date and time of training.
(2)   Location of training.
(3)   Subject of training.
(4)   Contents of training.
(5)   Length of training.
(6)   Names and signatures of attendees.
(7)   Names and signatures of instructors.

19　RECORD KEEPING

The Occupational Safety and Health Administration (OSHA) record-keeping system is mandatory for all establishments subject to the Occupational Safety and Health Act of 1970. The records required by OSHA (29 CFR 1904) are used to compile statistics on occupational safety and health; to study the causes and prevention of occupational injury and illness; and to inform employees about the work environment, its hazards, and their potential effects.

Under Part 1904, the records required must be kept according to a set of standardized rules. However, Section 1910.20, which provides more detailed information for studies of specific hazards, requires access to existing records but does not specify their form or content. The records open to inspection under Section 1910.20 include confidential medical information which OSHA must handle according to the procedures in Part 1913.10. Both 1904 and 1910.20 include paragraphs requiring that employees have access to certain records. The log and summary required in Part 1904 must be available to any employee. The standard gives employees and their representatives access to their own medical records and to company records of exposure to hazardous conditions in the workplace. The responsibility for providing employee access to this information rests with the employer. In addition, OSHA can gain access to the same information, including medical records, without the consent of employees.

The OSHA Laboratory Standard (29 CFR 1910.1450) requires employers of laboratories to establish and maintain for every employee an accurate record of any actions to monitor employee exposures and any medical consultation and examinations including tests and written opinions. These records must be maintained in accordance with 29 CFR 1910.20.

PURPOSE OF SECTION 1910.20

On September 29, 1988, OSHA issued a final rule on access to employee exposure and medical records. This final rule incorporates essentially the same provisions as those established in 1980, with the following major exceptions: first aid and medical records of short-term employees are exempt from record retention requirements; the microfilm storage of all employee X-rays except chest X-rays is permitted; employer trade secrets are provided additional protection and are made to conform with the Hazard Communication Standard (29 CFR 1910.1200); union representatives are required to show an occupational health need when seeking unconsented access to employee exposure records; and no industries are treated separately with respect to trade secret disclosure.

The purpose of Section 1910.20 is to give employees and their designated representatives, as well as OSHA, a right of access to relevant exposure and medical records. Employers are responsible for compliance with this section but may designate the physician or other health care personnel in charge of the records to carry out its provisions.

PRESERVATION OF RECORDS

The following is a summary of requirements of preservation of records under 29 CFR 1910.20:

(1) Employee medical records must be maintained for at least the duration of employment plus 30 years.

(2) The medical records of employees who have worked less than 1 year do not need to be kept beyond the term of employment if they are provided to the employee upon leaving.

(3) Employee exposure records must be kept for 30 years.

(4) Background data to employee exposure monitoring, e.g., laboratory reports and worksheets, need only be maintained for 1 year.

(5) Material Safety Data Sheets or other records identifying the toxic substance or harmful physical agent need not be retained as long as a record of the chemical substance and where and when it was used is kept for 30 years.

(6) Analysis using exposure or medical records must be kept for at least 30 years.

(7) Records can be maintained in any form as long as they are preserved and retrievable, except that chest X-ray films must be preserved in their original state.

ACCESS TO RECORDS

The following is a summary on employee access to records:

(1) Employees must be given access to records within 15 days of their request and at a reasonable time, place, and manner.

(2) An employee who requests a copy of a record may be given a free copy, allowed to use available copying facilities, or allowed to borrow the record to make a copy.

(3) Employers may charge reasonable administrative costs to answer subsequent requests for the same record.

(4) Employees must have access to all records relevant to their exposure and to medical records of which the employee is the subject.

(5) Employees must be allowed access to analysis using their exposure or medical records unless it is impossible to remove personal identifiers from the analysis.

(6) OSHA must be allowed access upon request to employee exposure or medical records.

(7) If OSHA wants access to medical records which identify the employees involved, it must present, and the employer must post, a written access order.

(8) OSHA has recognized that, with regard to medical records, the nature of the information may warrant discretion and has described certain conditions to be considered as valid exceptions.

TRADE SECRETS

Trade secret data disclosing the manufacturing process or percentage composition of a chemical may be deleted from requested records as long as the deletion is noted and alternative information is provided where necessary to identify where and when exposure occurred. However, when a physician determines that a medical emergency exists and that the chemical identity of a toxic substance is needed for treatment, the employer must provide that information immediately regardless of any previous written requests for confidentiality. The employer may require a written statement of confidentiality from the employee and/or the medical authority when such information is released, as soon as circumstances permit. There are also certain non-emergency situations which would require the employer to disclose specific chemical identities of toxic substances. When an employer makes a trade secret claim, it must be made no later than at the time the information is given to the Assistant Secretary of Labor so that the necessary protection can be implemented.

EMPLOYEE INFORMATION

Employees exposed to hazardous chemicals or harmful physical agents must be informed annually of their right of access to the records, the location of such records, and the person responsible for providing access. Employers must also make a copy of the standard and its appendices readily available to employees and distribute any information provided by OSHA.

TRANSFER OF RECORDS

When an employer goes out of business without a successor or when the retention period has run out, records cannot simply be destroyed. The steps to be taken include notification of both employees and the National Institute for Occupational Safety and Health (NIOSH) to allow information of value to be claimed by concerned parties.

BIBLIOGRAPHY

American National Standard for Emergency Eyewash and Shower Equipment: Standard Z358.1-1990, American National Standards Institute, New York, 1990.

American National Standard for Hazardous Industrial Chemicals — Precautionary Labeling: Standard Z129.1-1988, American National Standards Institute, New York, 1988.

American National Standard for Occupational and Educational Eye and Face Protection: Standard Z87.1-1989, American National Standards Institute, New York, 1989.

American National Standard for Practices for Respiratory Protection: Standard Z88.2-1980, American National Standards Institute, New York, 1980.

Annual Report on Carcinogens, 8th ed., DHHS(NTP), Summary from National Toxicology Program, Public Information Office, Research Triangle Park, NC, 1993.

Best's Safety Directory, A.M. Best, Oldwick, NJ, 1993.

Beyler, R.E. and Myers, V.K., What every chemist should know about teratogens, *J. Chem. Ed.,* 59, 759, 1982.

Braker, W. and Mossman, A.L., *Gas Data Book,* 6th ed., Matheson Gas Products, East Rutherford, NJ, 1980.

Bretherick, L., *Handbook of Reactive Chemical Hazards,* 4th ed., Butterworths, London, 1990.

Bretherick, L., *Hazards in the Chemical Laboratory,* 4th ed., Royal Society of Chemistry, London, 1986.

Carcinogens — Regulation and Control: A Management Guide to Carcinogens, NIOSH, DHEW Publication No. 77-205, Government Printing Office, Washington, D.C., 1977.

Carcinogens — Working with Carcinogens: A Guide to Good Health Practices, NIOSH, DHEW Publication No. 77-206, Government Printing Office, Washington, D.C., 1977.

Chemical and Engineering News: Safety Notes Index, 1976–1989, American Chemical Society, Washington, D.C.

Clansky, R.B., Ed., *Chemical Guide to the OSHA Hazard Communication Standard,* Roytech, Burlingame, CA, 1992.

Clansky, R.B., Ed., *Suspect Chemicals Sourcebook: Guide to Industrial Chemicals Covered Under Major Federal Regulatory and Advisory Programs,* Roytech, Burlingame, CA, 1992.

Clayton, G.D. and Clayton, F.E., Eds., *Patty's Industrial Hygiene and Toxicology,* 4th ed., Wiley-Interscience, New York, 1991.

Committee on Chemical Safety, *Design of Safe Chemical Laboratories: Suggested References,* 2nd ed., American Chemical Society, Washington, D.C., 1991.

Committee on Chemical Safety, *Guidelines for Authors of Laboratory Manuals,* 2nd ed., American Chemical Society, Washington, D.C., 1990.

Committee on Chemical Safety, *Safety in Academic Chemistry Laboratories,* 5th ed., American Chemical Society, Washington, D.C., 1990.

Committee on Chemists with Disabilities, *Teaching Chemistry to Students with Disabilities,* American Chemical Society, Washington, D.C., 1993.

Department of Government Relations and Science Policy, *Hazardous Waste Management Information Pamphlet,* American Chemical Society, Washington, D.C., 1992.

Department of Government Relations and Science Policy, *Less Is Better: Laboratory Chemical Management for Waste Reduction,* American Chemical Society, Washington, D.C., 1985.

Department of Government Relations and Science Policy, *Waste Management Manual for Laboratory Personnel,* American Chemical Society, Washington, D.C., 1990.

Diberardinis, L.J., Blum, J.S., Gatwood, G.T., Seth, A.K., First, M.W., and Groden, E.F., *Guidelines for Laboratory Design: Health and Safety Considerations,* 2nd ed., Wiley-Interscience, New York, 1992.

Fire Protection for Laboratories Using Chemicals: NFPA Standard 45, National Fire Protection Association, Quincy, MA, 1991.

Fire Protection Guide to Hazardous Materials [contains NFPA 325M, NFPA 49, NFPA 491M, and NFPA 704], 10th ed., National Fire Protection Association, Quincy, MA, 1991.

Flammable and Combustible Liquids Code: NFPA Standard 30, National Fire Protection Association, Quincy, MA, 1990.

Fawcett, H.H., Ed., *Hazardous and Toxic Materials: Safe Handling and Disposal,* 2nd ed., Wiley-Interscience, New York, 1988.

Freeman, H., Ed., *Hazardous Waste Minimization,* McGraw-Hill, New York, 1990.

Fuscaldo, A.A., Erlick, B.J., and Hindman, B., *Laboratory Safety: Theory and Practice,* Academic Press, New York, 1980.

Gerlovich, J.A. and Downs, G.E., *Better Science Through Safety,* Iowa State University Press, Ames, IA, 1984.

Gosselin, R.E., Ed., *Clinical Toxicology of Commercial Products, Acute Poisoning,* 5th ed., Williams and Wilkins, Baltimore, 1984.

Green, M.E. and Turk, A., *Safety in Working with Chemicals,* Macmillan, New York, 1978.

Handbook of Compressed Gases, 3rd ed., Compressed Gas Association, New York, 1990.

Hazardous Materials Information Center, *Hazard Classification Systems: Comparative Guide to Definitions and Labels,* Inter/Face, Middletown, CT, 1986.

IARC Monographs on the Evaluation of Carcinogenic Risk of Chemicals to Humans, International Agency for Research on Cancer, Lyon, France, 1992.

Kamrin, M.A., *Toxicology: A Primer on Toxicology Principles and Applications,* Lewis Publishers, Chelsea, MI, 1988.

Kaufman, J.A., *Waste Disposal in Academic Institutions,* Lewis Publishers, Boca Raton, FL, 1990.

Lewis, R.J., Ed., *Hawley's Condensed Chemical Dictionary,* 12th ed., Van Nostrand Reinhold, New York, 1992.

Lewis, R.L., *Carcinogenically Active Chemicals: a Reference Guide,* Van Nostrand Reinhold, New York, 1991.

Lunn, G. and Sansone, E.B., *Destruction of Hazardous Chemicals in the Laboratory,* Wiley-Interscience, New York, 1990.

Manual of Hazardous Chemical Reactions: NFPA Manual 491M, National Fire Protection Association, Quincy, MA, 1991.

Mayo, D.D.W., Butcher, S.S., Pike, R.M., Foote, C.M., Hotham, J.R., and Page, D.S., *Microscale Organic Laboratory,* John Wiley & Sons, New York, 1986.

Merck Index: an Encyclopedia of Chemicals, Drugs, and Biologicals, 11th ed., Merck, Rahway, NJ, 1989.

Mikell, W.G. and Drinkard, W.C., Good practices for hood use, *J. Chem. Ed.,* 61, A13, 1984.

Mikell, W.G. and Fuller, F.H., Good hood practices for safe hood operation, *J. Chem. Ed.,* 65, A36, 1988.

Mikell, W.G. and Hobbs, L.R., Laboratory hood studies, *J. Chem. Ed.,* 58, A165, 1981.

National Electrical Code, Vol. 5 of the National Fire Codes, National Fire Protection Association, Quincy, MA, (updated periodically).

NIH Guidelines for the Laboratory Use of Chemical Carcinogens, NIH, DHHS, Government Printing Office, Washington, D.C., 1981.

NIOSH Pocket Guide to Chemical Hazards, NIOSH Publication No. 90-117, Government Printing Office, Washington, D.C., 1990.

Occupational Health Guidelines for Chemical Hazards, 3 vols., NIOSH Publication No. 81- 123, Government Printing Office, Washington, D.C., 1981.

Phifer, R.W. and McTigue, W.R., Jr., *Handbook of Hazardous Waste Management for Small Quantity Generators,* Lewis Publishers, Chelsea, MI, 1985.

Pipitone, D.A., Ed., *Safe Storage of Laboratory Chemicals,* 2nd ed., Wiley-Interscience, New York, NY, 1991.

Pitt, M.J. and Pitt, E., *Handbook of Laboratory Waste Disposal: A Practical Manual,* Halsted, New York, 1985.

Pitt, M.J., Please do not touch — some thoughts on temporary labels in the laboratory, *J. Chem. Ed.,* 61, A231, 1984.

Prudent Practices for Handling Hazardous Chemicals in Laboratories, National Academy of Sciences, Washington, D.C., 1981.

Prudent Practices for the Disposal of Chemicals from the Laboratory, National Academy of Sciences, Washington, D.C., 1983.

Registry of Toxic Effects of Chemical Substances, NIOSH, DHHS, Government Printing Office, Washington, D.C., (updated periodically).

Reproductive Health Hazards in the Workplace, Office of Technology Assessment, Government Printing Office, Washington, D.C., 1985.

Safety in the Chemical Laboratory, 4 vols., reprints from *Journal of Chemical Education,* Springfield, PA, 1964-1970.

Shane, B.S., Human reproductive hazards: evaluation and chemical etiology, *Environ. Sci. Technol.,* 23(10), 1187, 1989.

Shepard, T.H., *Catalogue of Teratogenic Agents,* 7th ed., Johns Hopkins University Press, Baltimore, MD, 1992.

Sittig, M., *Handbook of Toxic and Hazardous Chemicals and Carcinogens,* 3rd ed., Noyes Data Corporation, Park Ridge, NJ, 1992.

Stricoff, R.S. and Walters, D.B., *Laboratory Health and Safety Handbook: a Guide for the Preparation of A Chemical Hygiene Plan,* Wiley-Interscience, New York, 1990.

Threshold Limit Values for Chemical Substances and Physical Agents and Biological Exposure Indices, American Conference of Governmental Industrial Hygienists, Cincinnati, OH, 1993.

Walters, D.B., Ed., *Safe Handling of Chemical Carcinogens, Mutagens, Teratogens, and Highly Toxic Substances,* Ann Arbor Science, Ann Arbor, MI, 1980.

Young, J.A., Ed., *Improving Safety in the Chemical Laboratory: a Practical Guide,* 2nd ed., Wiley-Interscience, New York, 1991.

Young, J.A., Kingsley, W.R., and Wahl, G.H., *Developing a Chemical Hygiene Plan,* American Chemical Society, Washington, D.C., 1990.

States with OSHA-Approved Plans

States	Initial Approval Since	Operational Status Agreement[a]	On-Site Different Standards[b]	On-Shore Consultation Agreement[c]	Maritime Coverage
Alaska	07/31/73		Yes	Yes	
Arizona	10/29/74				
California	04/24/73	Yes	Yes	Yes	Yes
Connecticut	10/02/73			Yes	
Hawaii	12/28/73		Yes	Yes	
Indiana	02/25/74				
Iowa	07/12/73			Yes	
Kentucky	07/23/73				
Maryland	06/28/73			Yes	
Michigan	09/24/73	Yes	Yes	Yes	
Minnesota	05/29/73			Yes	Yes
Nevada	12/04/73	Yes			
New Mexico	12/04/75	Yes			
New York	06/01/84			Yes	
North Carolina	01/26/73	Yes		Yes	
Oregon	12/22/72	Yes	Yes	Yes	Yes
Puerto Rico	08/15/77	Yes			
South Carolina	11/30/72			Yes	
Tennessee	06/28/73			Yes	
Utah	01/04/73			Yes	
Vermont	10/01/73	Yes		Yes	Yes
Virginia	09/23/76			Yes	
Virgin Island	08/31/73				
Washington	01/19/73	Yes	Yes		Yes
Wyoming	04/25/74			Yes	

[a] Concurrent federal jurisdiction suspended.
[b] State standard frequently not identical to the federal standard.
[c] On-site consultation available in these states.

Offices of State Programs

State	Office of State Program	Phone No.
Alaska	Alaska Department of Labor P.O. Box 2149 Juneau, AK 99802	(907) 465-2700
Arizona	Industrial Commission of Arizona 800 West Washington Phoenix, AZ 85007	(602) 542-2795
California	California Department of Industrial Relations 395 Oyster Point Boulevard South San Francisco, CA 94080	(415) 737-2960
Connecticut	Connecticut Department of Labor 200 Folly Brook Boulevard Wethersfield, CT 06109	(203) 566-5123
Hawaii	Hawaii Department of Labor and Industrial Relations 830 Punchbowl Street Honolulu, HI 96813	(808) 548-3150
Indiana	Indiana Department of Labor 100 North Senate Avenue Indianapolis, IN 46204	(317) 232-2665
Iowa	Iowa Division of Labor Services 1000 East Grand Avenue Des Moines, IA 50319	(515) 281-3447
Kentucky	Kentucky Labor Cabinet U.S. Highway 127 South Frankfort, KY 40601	(502) 564-3070
Maryland	Maryland Division of Labor and Industry 501 St. Paul Place Baltimore, MD 21202	(301) 333-4179
Michigan	Michigan Department of Labor 201 North Washington Square Lansing, MI 48933	(517) 373-9600
Minnesota	Minnesota Department of Labor and Industry 443 Lafayette Road St. Paul, MN 55155	(612) 296-2342
Nevada	Nevada Department of Industrial Relations 1370 South Curry Street Carson City, NV 89701	(702) 885-5240

New Mexico	New Mexico Health and Environment Department 1190 St. Francis Drive Santa Fe, NM 87503	(505) 827-2850
New York	New York Department of Labor One Main Street Brooklyn, NY 11201	(518) 457-3518
North Carolina	North Carolina Department of Labor 4 West Edenton Street Raleigh, NC 27603	(919) 733-7166
Oregon	Oregon Occupational Safety and Health Division Labor and Industries Building Salem, OR 97310	(503) 378-3272
Puerto Rico	Puerto Rico Department of Labor and Human Resources 505 Munoz Rivera Avenue Hato Rey, Puerto Rico 00918	(809) 754-2119
South Carolina	South Carolina Department of Labor 3600 Forest Drive Columbia, SC 29211	(803) 734-9594
Tennessee	Tennessee Department of Labor 501 Union Building Nashville, TN 37219	(615) 741-2582
Utah	Utah Occupational Safety and Health 160 East 300 South Salt Lake City, UT 84110	(801) 530-6900
Vermont	Vermont Department of Labor and Industry 120 State Street Montpelier, VT 05602	(802) 828-2765
Virgin Islands	Virgin Islands Department of Labor P.O. Box 890 St. Croix, Virgin Islands 00820	(809) 773-1994
Virginia	Virginia Department of Labor and Industry P.O. Box 12064 Richmond, VA 23241	(804) 786-2376
Washington	Washington Department of Labor and Industries General Administration Building Olympia, WA 98504	(206) 753-6307
Wyoming	Wyoming Occupational Health and Safety 122 West 25th Street Cheyenne, WY 82002	(307) 777-7786

APPENDIX B

OSHA Laboratory Standard
Rules and Regulations
Taken from the Federal Register, Vol. 55, No. 21
Wednesday, January 31, 1990

engaged in the laboratory use of hazardous chemicals as defined below.

(2) Where this section applies, it shall supersede, for laboratories, the requirements of all other OSHA health standards in 29 CFR part 1910, subpart Z, except as follows:

(i) For any OSHA health standard, only the requirement to limit employee exposure to the specific permissible exposure limit shall apply for laboratories, unless that particular standard states otherwise or unless the conditions of paragraph (a)(2)(iii) of this section apply.

(ii) Prohibition of eye and skin contact where specified by any OSHA health standard shall be observed.

(iii) Where the action level (or in the absence of an action level, the permissible exposure limit) is routinely exceeded for an OSHA regulated substance with exposure monitoring and medical surveillance requirements, paragraphs (d) and (g)(1)(ii) of this section shall apply.

(3) This section shall not apply to:

(i) Uses of hazardous chemicals which do not meet the definition of laboratory use, and in such cases, the employer shall comply with the relevant standard in 29 CFR part 1910, subpart Z, even if such use occurs in a laboratory.

(ii) Laboratory uses of hazardous chemicals which provide no potential for employee exposure. Examples of such conditions might include:

(A) Procedures using chemically-impregnated test media such as Dip-and-Read tests where a reagent strip is dipped into the specimen to be tested and the results are interpreted by comparing the color reaction to a color chart supplied by the manufacturer of the test strip; and

(B) Commercially prepared kits such as those used in performing pregnancy tests in which all of the reagents needed to conduct the test are contained in the kit.

(b) *Definitions*—

"*Action level*" means a concentration designated in 29 CFR part 1910 for a specific substance, calculated as an eight (8)-hour time-weighted average, which initiates certain required activities such as exposure monitoring and medical surveillance.

"*Assistant Secretary*" means the Assistant Secretary of Labor for Occupational Safety and Health, U.S. Department of Labor, or designee.

"*Carcinogen*" (see "select carcinogen").

"*Chemical Hygiene Officer*" means an employee who is designated by the employer, and who is qualified by training or experience, to provide technical guidance in the development

and implementation of the provisions of the Chemical Hygiene Plan. This definition is not intended to place limitations on the position description or job classification that the designated indvidual shall hold within the employer's organizational structure.

"*Chemical Hygiene Plan*" means a written program developed and implemented by the employer which sets forth procedures, equipment, personal protective equipment and work practices that (i) are capable of protecting employees from the health hazards presented by hazardous chemicals used in that particular workplace and (ii) meets the requirements of paragraph (e) of this section.

"*Combustible liquid*" means any liquid having a flashpoint at or above 100 °F (37.8 °C), but below 200 °F (93.3 °C), except any mixture having components with flashpoints of 200 °F (93.3 °C), or higher, the total volume of which make up 99 percent or more of the total volume of the mixture.

"*Compressed gas*" means:

(i) A gas or mixture of gases having, in a container, an absolute pressure exceeding 40 psi at 70 °F (21.1 °C); or

(ii) A gas or mixture of gases having, in a container, an absolute pressure exceeding 104 psi at 130 °F (54.4 °C) regardless of the pressure at 70 °F (21.1 °C); or

(iii) A liquid having a vapor pressure exceeding 40 psi at 100 °F (37.8 °C) as determined by ASTM D–323–72.

"*Designated area*" means an area which may be used for work with "select carcinogens," reproductive toxins or substances which have a high degree of acute toxicity. A designated area may be the entire laboratory, an area of a laboratory or a device such as a laboratory hood.

"*Emergency*" means any occurrence such as, but not limited to, equipment failure, rupture of containers or failure of control equipment which results in an uncontrolled release of a hazardous chemical into the workplace.

"*Employee*" means an individual employed in a laboratory workplace who may be exposed to hazardous chemicals in the course of his or her assignments.

"*Explosive*" means a chemical that causes a sudden, almost instantaneous release of pressure, gas, and heat when subjected to sudden shock, pressure, or high temperature.

"*Flammable*" means a chemical that falls into one of the following categories:

(i) "*Aerosol, flammable*" means an aerosol that, when tested by the method described in 16 CFR 1500.45, yields a

PART 1910—OCCUPATIONAL SAFETY AND HEALTH STANDARDS

1. The authority citation for part 1910, subpart Z is amended by adding the following citation at the end. (Citation which precedes asterisk indicates general rulemaking authority.)

Authority: Secs. 6 and 8, Occupational Safety and Health Act, 29 U.S.C. 655, 657; Secretary of Labor's Orders Nos. 12–71 (36 FR 8754), 8–76 (41 FR 25059), or 9–83 (48 FR 35736), as applicable; and 29 CFR part 1911. * * * Section 1910.1450 is also issued under sec. 6(b), 8(c) and 8(g)(2), Pub. L. 91–596, 84 Stat. 1593, 1599, 1600; 29 U.S.C. 655, 657.

2. Section 1910.1450 is added to subpart Z, part 1910 to read as follows:

§ 191.1450 Occupational exposure to hazardous chemicals in laboratories.

(a) *Scope and application.* (1) This section shall apply to all employers flame protection exceeding 18 inches at full valve opening, or a flashback (a flame extending back to the valve) at any degree of valve opening;

(ii) "*Gas, flammable*" means:

(A) A gas that, at ambient temperature and pressure, forms a flammable mixture with air at a concentration of 13 percent by volume or less; or

(B) A gas that, at ambient temperature and pressure, forms a range of flammable mixtures with air wider than 12 percent by volume, regardless of the lower limit.

(iii) "*Liquid, flammable*" means any liquid having a flashpoint below 100 °F (37.8 °C), except any mixture having components with flashpoints of 100 °F (37.8 °C) or higher, the total of which make up 99 percent or more of the total volume of the mixture.

(iv) "*Solid, flammable*" means a solid, other than a blasting agent or explosive as defined in § 1910.109(a), that is liable to cause fire through friction, absorption of moisture, spontaneous chemical change, or retained heat from manufacturing or processing, or which can be ignited readily and when ignited burns so vigorously and persistently as to create a serious hazard. A chemical shall be considered to be a flammable solid if, when tested by the method described in 16 CFR 1500.44, it ignites and burns with a self-sustained flame at a rate greater than one-tenth of an inch per second along its major axis.

"*Flashpoint*" means the minimum temperature at which a liquid gives off a vapor in sufficient concentration to ignite when tested as follows:

(i) Tagliabue Closed Tester (See American National Standard Method of Test for Flash Point by Tag Closed Tester, Z11.24–1979 (ASTM D 56–79)]-for liquids with a viscosity of less than 45 Saybolt Universal Seconds (SUS) at 100 °F (37.8 °C), that do not contain suspended solids and do not have a tendency to form a surface film under test; or

(ii) Pensky-Martens Closed Tester (see American National Standard Method of Test for Flash Point by Pensky-Martens Closed Tester, Z11.7–1979 (ASTM D 93–79)]-for liquids with a viscosity equal to or greater than 45 SUS at 100 °F (37.8 °C), or that contain suspended solids, or that have a tendency to form a surface film under test; or

(iii) Setaflash Closed Tester (see American National Standard Method of Test for Flash Point by Setaflash Closed Tester (ASTM D 3278–78)].

Organic peroxides, which undergo autoaccelerating thermal decomposition, are excluded from any of the flashpoint determination methods specified above.

"*Hazardous chemical*" means a chemical for which there is statistically significant evidence based on at least one study conducted in accordance with established scientific principles that acute or chronic health effects may occur in exposed employees. The term "health hazard" includes chemicals which are carcinogens, toxic or highly toxic agents, reproductive toxins, irritants, corrosives, sensitizers, hepatotoxins, nephrotoxins, neurotoxins, agents which act on the hematopoietic systems, and agents which damage the lungs, skin, eyes, or mucous membranes.

Appendices A and B of the Hazard Communication Standard (29 CFR 1910.1200) provide further guidance in defining the scope of health hazards and determining whether or not a chemical is to be considered hazardous for purposes of this standard.

"*Laboratory*" means a facility where the "laboratory use of hazardous chemicals" occurs. It is a workplace where relatively small quantities of hazardous chemicals are used on a non-production basis.

"*Laboratory scale*" means work with substances in which the containers used for reactions, transfers, and other handling of substances are designed to be easily and safely manipulated by one person. "Laboratory scale" excludes those workplaces whose function is to produce commercial quantities of materials.

"*Laboratory-type hood*" means a device located in a laboratory, enclosure on five sides with a moveable sash or fixed partial enclosed on the remaining side: constructed and maintained to draw air from the laboratory and to prevent or minimize the escape of air contaminants into the laboratory; and allows chemical manipulations to be conducted in the enclosure without insertion of any portion of the employee's body other than hands and arms.

Walk-in hoods with adjustable sashes meet the above definition provided that the sashes are adjusted during use so that the airflow and the exhaust of air contaminants are not compromised and employees do not work inside the enclosure during the release of airborne hazardous chemicals.

"*Laboratory use of hazardous chemicals*" means handling or use of such chemicals in which all of the following conditions are met:

(i) Chemical manipulations are carried out on a "laboratory scale;"

(ii) Multiple chemical procedures or chemicals are used;

(iii) The procedures involved are not part of a production process, nor in any way simulate a production process; and

(iv) "Protective laboratory practices and equipment" are available and in common use to minimize the potential for employee exposure to hazardous chemicals.

"*Medical consultation*" means a consultation which takes place between an employee and a licensed physician for the purpose of determining what medical examinations or procedures, if any, are appropriate in cases where a significant exposure to a hazardous chemical may have taken place.

"*Organic peroxide*" means an organic compound that contains the bivalent $-O-O-$ structure and which may be considered to be a structural derivative of hydrogen peroxide where one or both of the hydrogen atoms has been replaced by an organic radical.

"*Oxidizer*" means a chemical other than a blasting agent or explosive as defined in § 1910.109(a), that initiates or promotes combustion in other materials, thereby causing fire either of itself or through the release of oxygen or other gases.

"*Physical hazard*" means a chemical for which there is scientifically valid evidence that it is a combustible liquid, a compressed gas, explosive, flammable, an organic peroxide, an oxidizer, pyrophoric, unstable (reactive) or water-reactive.

"*Protective laboratory practices and equipment*" means those laboratory procedures, practices and equipment accepted by laboratory health and safety experts as effective, or that the employer can show to be effective, in minimizing the potential for employee exposure to hazardous chemicals.

"*Reproductive toxins*" means chemicals which affect the reproductive capabilities including chromosomal damage (mutations) and effects on fetuses (teratogenesis)

"*Select carcinogen*" means any substance which meets one of the following criteria:

(i) It is regulated by OSHA as a carcinogen; or

(ii) It is listed under the category, "known to be carcinogens," in the Annual Report on Carcinogens published by the National Toxicology Program (NTP) (latest edition); or

(iii) It is listed under Group 1 ("carcinogenic to humans") by the International Agency for Research on Cancer Monographs (IARC) (latest editions); or

(iv) It is listed in either Group 2A or 2B by IARC or under the category, "reasonably anticipated to be carcinogens" by NTP, and causes statistically significant tumor incidence in experimental animals in accordance with any of the following criteria:

(A) After inhalation exposure of 6–7 hours per day, 5 days per week, for a significant portion of a lifetime to dosages of less than 10 mg/m^3;

(B) After repeated skin application of less than 300 (mg/kg of body weight) per week; or

(C) After oral dosages of less than 50 mg/kg of body weight per day.

"*Unstable (reactive)*" means a chemical which is the pure state, or as produced or transported, will vigorously polymerize, decompose, condense, or will become self-reactive under conditions of shocks, pressure or temperature.

"*Water-reactive*" means a chemical that reacts with water to release a gas that is either flammable or presents a health hazard.

(c) *Permissible exposure limits.* For laboratory uses of OSHA regulated substances, the employer shall assure that laboratory employees' exposures to such substances do not exceed the permissible exposure limits specified in 29 CFR part 1910, subpart Z.

(d) *Employee exposure determination*—(1) *Initial monitoring.* The employer shall measure the employee's exposure to any substance regulated by a standard which requires monitoring if there is reason to believe that exposure levels for that substance routinely exceed the action level (or in the absence of an action level, the PEL).

(2) *Periodic monitoring.* If the initial monitoring prescribed by paragraph (d)(1) of this section discloses employee exposure over the action level (or in the absence of an action level, the PEL), the employer shall immediately comply with the exposure monitoring provisions of the relevant standard.

(3) *Termination of monitoring.* Monitoring may be terminated in accordance with the relevant standard.

(4) *Employee notification of monitoring results.* The employer shall, within 15 working days after the receipt of any monitoring results, notify the employee of these results in writing either individually or by posting results in an appropriate location that is accessible to employees.

(e) *Chemical hygiene plan—General.* (Appendix A of this section is non-mandatory but provides guidance to assist employers in the development of the Chemical Hygiene Plan.) (1) Where hazardous chemicals as defined by this standard are used in the workplace, the employer shall develop and carry out the provisions of a written Chemical Hygiene Plan which is:

(i) Capable of protecting employees from health hazards associated with hazardous chemicals in that laboratory and

(ii) Capable of keeping exposures below the limits specified in paragraph (c) of this section.

(2) The Chemical Hygiene Plan shall be readily available to employees, employee representatives and, upon request, to the Assistant Secretary.

(3) The Chemical Hygiene Plan shall include each of the following elements and shall indicate specific measures that the employer will take to ensure laboratory employee protection:

(i) Standard operating procedures relevant to safety and health considerations to be followed when laboratory work involves the use of hazardous chemicals;

(ii) Criteria that the employer will use to determine and implement control measures to reduce employee exposure to hazardous chemicals including engineering controls, the use of personal protective equipment and hygiene practices; particular attention shall be given to the selection of control measures for chemicals that are known to be extremely hazardous;

(iii) A requirement that fume hoods and other protective equipment are functioning properly and specific measures that shall be taken to ensure proper and adequate performance of such equipment;

(iv) Provisions for employee information and training as prescribed in paragraph (f) of this section;

(v) The circumstances under which a particular laboratory operation, procedure or activity shall require prior approval from the employer or the employer's designee before implementation;

(vi) Provisions for medical consultation and medical examinations in accordance with paragraph (g) of this section;

(vii) Designation of personnel responsible for implementation of the Chemical Hygiene Plan including the assignment of a Chemical Hygiene Officer and, if appropriate, establishment of a Chemical Hygiene Committee; and

(viii) Provisions for additional employee protection for work with particularly hazardous substances. These include "select carcinogens," reproductive toxins and substances which have a high degree of acute toxicity. Specific consideration shall be given to the following provisions which shall be included where appropriate:

(A) Establishment of a designated area;

(B) Use of containment devices such as fume hoods or glove boxes;

(C) Procedures for safe removal of contaminated waste; and

(D) Decontamination procedures.

(4) The employer shall review and evaluate the effectiveness of the Chemical Hygiene Plan at least annually and update it as necessary.

(f) *Employee information and training.* (1) The employer shall provide employees with information and training to ensure that they are apprised of the hazards of chemicals present in their work area.

(2) Such information shall be provided at the time of an employee's initial assignment to a work area where hazardous chemicals are present and prior to assignments involving new exposure situations. The frequency of refresher information and training shall be determined by the employer.

(3) *Information.* Employees shall be informed of:

(i) The contents of this standard and its appendices which shall be made available to employees;

(ii) The location and availability of the employer's Chemical Hygiene Plan;

(iii) The permissible exposure limits for OSHA regulated substances or recommended exposure limits for other hazardous chemicals where there is no applicable OSHA standard;

(iv) Signs and symptoms associated with exposures to hazardous chemicals used in the laboratory; and

(v) The location and availability of known reference material on the hazards, safe handling, storage and disposal of hazardous chemicals found in the laboratory including, but not limited to, Material Safety Data Sheets received from the chemical supplier.

(4) *Training.* (i) Employee training shall include:

(A) Methods and observations that may be used to detect the presence or release of a hazardous chemical (such as monitoring conducted by the employer, continuous monitoring devices, visual appearance or odor of hazardous chemicals when being released, etc.);

(B) The physical and health hazards of chemicals in the work area; and

(C) The measures employees can take to protect themselves from these hazards, including specific procedures the employer has implemented to protect employees from exposure to hazardous chemicals, such as appropriate work practices, emergency procedures, and personal protective equipment to be used.

(ii) The employee shall be trained on the applicable details of the employer's written Chemical Hygiene Plan.

(g) *Medical consultation and medical examinations.* (1) The employer shall provide all employees who work with hazardous chemicals an opportunity to receive medical attention, including any follow-up examinations which the examining physician determines to be necessary, under the following circumstances:

(i) Whenever an employee develops signs or symptoms associated with a hazardous chemical to which the employee may have been exposed in the laboratory, the employee shall be provided an opportunity to receive an appropriate medical examination.

(ii) Where exposure monitoring reveals an exposure level routinely above the action level (or in the absence of an action level, the PEL) for an OSHA regulated substance for which there are exposure monitoring and medical surveillance requirements, medical surveillance shall be established for the affected employee as prescribed by the particular standard.

(iii) Whenever an event takes place in the work area such as a spill, leak, explosion or other occurrence resulting in the likelihood of a hazardous exposure, the affected employee shall be provided an opportunity for a medical consultation. Such consultation shall be for the purpose of determining the need for a medical examination.

(2) All medical examinations and consultations shall be performed by or under the direct supervision of a licensed physician and shall be provided without cost to the employee, without loss of pay and at a reasonable time and place.

(3) *Information provided to the physician.* The employer shall provide the following information to the physician:

(i) The identity of the hazardous chemical(s) to which the employee may have been exposed;

(ii) A description of the conditions under which the exposure occurred including quantitative exposure data, if available; and

(iii) A description of the signs and symptoms of exposure that the employee is experiencing, if any.

(4) *Physician's written opinion.* (i) For examination or consultation required under this standard, the employer shall obtain a written opinion from the examining physician which shall include the following:

(A) Any recommendation for further medical follow-up;

(B) The results of the medical examination and any associated tests;

(C) Any medical condition which may be revealed in the course of the examination which may place the employee at increased risk as a result of exposure to a hazardous chemical found in the workplace; and

(D) A statement that the employee has been informed by the physician of the results of the consultation or medical examination and any medical condition that may require further examination or treatment.

(ii) The written opinion shall not reveal specific findings of diagnoses unrelated to occupational exposure.

(h) *Hazard identification.* (1) With respect to labels and material safety data sheets:

(i) Employers shall ensure that labels on incoming containers of hazardous chemicals are not removed or defaced.

(ii) Employers shall maintain any material safety data sheets that are received with incoming shipments of hazardous chemicals, and ensure that they are readily accessible to laboratory employees.

(2) The following provisions shall apply to chemical substances developed in the laboratory:

(i) If the composition of the chemical substance which is produced exclusively for the laboratory's use is known, the employer shall determine if it is a hazardous chemical as defined in paragraph (b) of this section. If the chemical is determined to be hazardous, the employer shall provide appropriate training as required under paragraph (f) of this section.

(ii) If the chemical produced is a byproduct whose composition is not known, the employer shall assume that the substance is hazardous and shall implement paragraph (e) of this section.

(iii) If the chemical substance is produced for another user outside of the laboratory, the employer shall comply with the Hazard Communication Standard (29 CFR 1910.1200) including the requirements for preparation of material safety data sheets and labeling.

(i) *Use of respirators.* Where the use of respirators is necessary to maintain exposure below permissible exposure limits, the employer shall provide, at no cost to the employee, the proper respiratory equipment. Respirators shall be selected and used in accordance with the requirements of 29 CFR 1910.134.

(j) *Recordkeeping.* (1) The employer shall establish and maintain for each employee an accurate record of any measurements taken to monitor employee exposures and any medical consultation and examinations including tests or written opinions required by this standard.

(2) The employer shall assure that such records are kept, transferred, and made available in accordance with 29 CFR 1910.20.

(k) *Dates*—(1) *Effective date.* This section shall become effective May 1, 1990.

(2) *Start-up dates.* (i) Employers shall have developed and implemented a written Chemical Hygiene Plan no later than January 31, 1991.

(ii) Paragraph (a)(2) of this section shall not take effect until the employer has developed and implemented a written Chemical Hygiene Plan.

(l) *Appendices.* The information contained in the appendices is not intended, by itself, to create any additional obligations not otherwise imposed or to detract from any existing obligation.

Appendix A to § 1910.1450—National Research Council Recommendations Concerning Chemical Hygiene in Laboratories (Non-Mandatory)

Table of Contents

Foreword

Corresponding Sections of the Standard and This Appendix

A. *General Principles*

1. Minimize all Chemical Exposures
2. Avoid Underestimation of Risk
3. Provide Adequate Ventilation
4. Institute a Chemical Hygiene Program
5. Observe the PELs and TLVs

B. *Responsibilities*

1. Chief Executive Officer
2. Supervisor of Administrative Unit
3. Chemical Hygiene Officer
4. Laboratory Supervisor
5. Project Director
6. Laboratory Worker

C. *The Laboratory Facility*

1. Design
2. Maintenance
3. Usage
4. Ventilation

D. *Components of the Chemical Hygiene Plan*

1. Basic Rules and Procedures
2. Chemical Procurement, Distribution, and Storage
3. Environmental Monitoring
4. Housekeeping, Maintenance and Inspections
5. Medical Program
6. Personal Protective Apparel and Equipment
7. Records
8. Signs and Labels
9. Spills and Accidents
10. Training and Information
11. Waste Disposal

E. *General Procedures for Working With Chemicals*

1. General Rules for all Laboratory Work with Chemicals
2. Allergens and Embryotoxins

Federal Register / Vol. 55, No. 21 / Wednesday, January 31, 1990 / Rules and Regulations   3331

3. Chemicals of Moderate Chronic or High Acute Toxicity

4. Chemicals of High Chronic Toxicity

5. Animal Work with Chemicals of High Chronic Toxicity

F. *Safety Recommendations*

G. *Material Safety Data Sheets*

Foreword

As guidance for each employer's development of an appropriate laboratory Chemical Hygiene Plan, the following non-mandatory recommendations are provided. They were extracted from "Prudent Practices for Handling Hazardous Chemicals in Laboratories" (referred to below as "Prudent Practices"), which was published in 1981 by the National Research Council and is available from the National Academy Press, 2101 Constitution Ave., NW., Washington DC 20418.

"Prudent Practices" is cited because of its wide distribution and acceptance and because of its preparation by members of the laboratory community through the sponsorship of the National Research Council. However, none of the recommendations given here will modify any requirements of the laboratory standard. This Appendix merely presents pertinent recommendations from "Prudent Practices", organized into a form convenient for quick reference during operation of a laboratory facility and during development and application of a Chemical Hygiene Plan. Users of this appendix should consult "Prudent Practices" for a more extended presentation and justification for each recommendation.

"Prudent Practices" deals with both safety and chemical hazards while the laboratory standard is concerned primarily with chemical hazards. Therefore, only those recommendations directed primarily toward control of toxic exposures are cited in this appendix, with the term "chemical hygiene" being substituted for the word "safety". However, since conditions producing or threatening physical injury often pose toxic risks as well, page references concerning major categories of safety hazards in the laboratory are given in section F.

The recommendations from "Prudent Practices" have been paraphrased, combined, or otherwise reorganized, and headings have been added. However, their sense has not been changed.

Corresponding Sections of the Standard and this Appendix

The following table is given for the convenience of those who are developing a Chemical Hygiene Plan which will satisfy the requirements of paragraph (e) of the standard. It indicates those sections of this appendix which are most pertinent to each of the sections of paragraph (e) and related paragraphs.

Paragraph and topic in laboratory standard	Relevant appendix section
(e)(3)(i) Standard operating procedures for handling toxic chemicals.	C, D, E
(e)(3)(ii) Criteria to be used for implementation of measures to reduce exposures.	D
(e)(3)(iii) Fume hood performance.	C4b
(e)(3)(iv) Employee information and training (including emergency procedures).	D10, D9
(e)(3)(v) Requirements for prior approval of laboratory activities.	E2b, E4b
(e)(3)(vi) Medical consultation and medical examinations.	D5, E4f
(e)(3)(vii) Chemical hygiene responsibilities.	B
(e)(3)(viii) Special precautions for work with particularly hazardous substances.	E2, E3, E4

In this appendix, those recommendations directed primarily at administrators and supervisors are given in sections A–D. Those recommendations of primary concern to employees who are actually handling laboratory chemicals are given in section E. (Reference to page numbers in "Prudent Practices" are given in parentheses.)

A. General Principles for Work with Laboratory Chemicals

In addition to the more detailed recommendations listed below in sections B–E, "Prudent Practices" expresses certain general principles, including the following:

1. *It is prudent to minimize all chemical exposures.* Because few laboratory chemicals are without hazards, general precautions for handling all laboratory chemicals should be adopted, rather than specific guidelines for particular chemicals (2, 10). Skin contact with chemicals should be avoided as a cardinal rule (198).

2. *Avoid underestimation of risk.* Even for substances of no known significant hazard, exposure should be minimized; for work with substances which present special hazards, special precautions should be taken (10, 37, 38). One should assume that any mixture will be more toxic than its most toxic component (30, 103) and that all substances of unknown toxicity are toxic (3, 34).

3. *Provide adequate ventilation.* The best way to prevent exposure to airborne substances is to prevent their escape into the working atmosphere by use of hoods and other ventilation devices (32, 198).

4. *Institute a chemical hygiene program.* A mandatory chemical hygiene program designed to minimize exposures is needed: it should be a regular, continuing effort, not merely a standby or short-term activity (6, 11). Its recommendations should be followed in academic teaching laboratories as well as by full-time laboratory workers (13).

5. *Observe the PELs, TLVs.* The Permissible Exposure Limits of OSHA and the Threshold Limit Values of the American Conference of Governmental Industrial Hygienists should not be exceeded (13).

B. Chemical Hygiene Responsibilities

Responsibility for chemical hygiene rests at all levels (6, 11, 21) including the:

1. *Chief executive officer*, who has ultimate responsibility for chemical hygiene within the institution and must, with other administrators, provide continuing support for institutional chemical hygiene (7, 11).

2. *Supervisor of the department or other administrative unit*, who is responsible for chemical hygiene in that unit (7).

3. *Chemical hygiene officer(s)*, whose appointment is essential (7) and who must:

(a) Work with administrators and other employees to develop and implement appropriate chemical hygiene policies and practices (7);

(b) Monitor procurement, use, and disposal of chemicals used in the lab (8);

(c) See that appropriate audits are maintained (8);

(d) Help project directors develop precautions and adequate facilities (10);

(e) Know the current legal requirements concerning regulated substances (50); and

(f) Seek ways to improve the chemical hygiene program (8, 11).

4. *Laboratory supervisor*, who has overall responsibility for chemical hygiene in the laboratory (21) including responsibility to:

(a) Ensure that workers know and follow the chemical hygiene rules, that protective equipment is available and in working order, and that appropriate training has been provided (21, 22);

(b) Provide regular, formal chemical hygiene and housekeeping inspections including routine inspections of emergency equipment (21, 171);

(c) Know the current legal requirements concerning regulated substances (50, 231);

(d) Determine the required levels of protective apparel and equipment (156, 160, 162); and

(e) Ensure that facilities and training for use of any material being ordered are adequate (215).

5. *Project director or director of other specific operation*, who has primary responsibility for chemical hygiene procedures for that operation (7).

6. *Laboratory worker*, who is responsible for:

(a) Planning and conducting each operation in accordance with the institutional chemical hygiene procedures (7, 21, 22, 230); and

(b) Developing good personal chemical hygiene habits (22).

C. The Laboratory Facility

1. *Design.* The laboratory facility should have:

(a) An appropriate general ventilation system (see C4 below) with air intakes and exhausts located so as to avoid intake of contaminated air (194);

(b) Adequate, well-ventilated stockrooms/storerooms (218, 219);

(c) Laboratory hoods and sinks (12, 162);

(d) Other safety equipment including eyewash fountains and drench showers (162, 169); and

(e) Arrangements for waste disposal (12, 240).

2. *Maintenance.* Chemical-hygiene-related equipment (hoods, incinerator, etc.) should undergo continuing appraisal and be modified if inadequate (11, 12).

3. *Usage.* The work conducted (10) and its scale (12) must be appropriate to the physical facilities available and, especially, to the quality of ventilation (13).

4. *Ventilation—(a) General laboratory ventilation.* This system should: Provide a source of air for breathing and for input to local ventilation devices (199); it should not be relied on for protection from toxic substances released into the laboratory (198); ensure that laboratory air is continually replaced, preventing increase of air concentrations of toxic substances during the working day (194); direct air flow into the laboratory from non-laboratory areas and out to the exterior of the building (194).

(b) *Hoods.* A laboratory hood with 2.5 linear feet of hood space per person should be provided for every 2 workers if they spend most of their time working with chemicals (199); each hood should have a continuous monitoring device to allow convenient confirmation of adequate hood performance before use (200, 209). If this is not possible, work with substances of unknown toxicity should be avoided (13) or other types of local ventilation devices should be provided (199). See pp. 201–206 for a discussion of hood design, construction, and evaluation.

(c) *Other local ventilation devices.* Ventilated storage cabinets, canopy hoods, snorkels. etc. should be provided as needed (199). Each canopy hood and snorkel should have a separate exhaust duct (207).

(d) *Special ventilation areas.* Exhaust air from glove boxes and isolation rooms should be passed through scrubbers or other treatment before release into the regular exhaust system (208). Cold rooms and warm rooms should have provisions for rapid escape and for escape in the event of electrical failure (209).

(e) *Modifications.* Any alteration of the ventilation system should be made only if thorough testing indicates that worker protection from airborne toxic substances will continue to be adequate (12, 193, 204).

(f) *Performance.* Rate: 4–12 room air changes/hour is normally adequate general ventilation if local exhaust systems such as hoods are used as the primary method of control (194).

(g) *Quality.* General air flow should not be turbulent and should be relatively uniform throughout the laboratory, with no high velocity or static areas (194, 195); airflow into and within the hood should not be excessively turbulent (200); hood face velocity should be adequate (typically 60–100 lfm) (200, 204).

(h) *Evaluation.* Quality and quantity of ventilation should be evaluated on installation (202), regularly monitored (at least every 3 months) (6, 12, 14, 195), and reevaluated whenever a change in local ventilation devices is made (12, 195, 207). See pp. 195–198 for methods of evaluation and for calculation of estimated airborne contaminant concentrations.

D. Components of the Chemical Hygiene Plan

1. Basic Rules and Procedures (Recommendations for these are given in section E. below)

2. Chemical Procurement, Distribution, and Storage

(a) *Procurement.* Before a substance is received, information on proper handling, storage, and disposal should be known to those who will be involved (215, 216). No container should be accepted without an adequate identifying label (216). Preferably, all substances should be received in a central location (216).

(b) *Stockrooms/storerooms.* Toxic substances should be segregated in a well-identified area with local exhaust ventilation (221). Chemicals which are highly toxic (227) or other chemicals whose containers have been opened should be in unbreakable secondary containers (219). Stored chemicals should be examined periodically (at least annually) for replacement, deterioration, and container integrity (218–19).

Stockrooms/storerooms should not be used as preparation or repackaging areas, should be open during normal working hours, and should be controlled by one person (219).

(c) *Distribution.* When chemicals are hand carried, the container should be placed in an outside container or bucket. Freight-only elevators should be used if possible (223).

(d) *Laboratory storage.* Amounts permitted should be as small as practical. Storage on bench tops and in hoods is inadvisable. Exposure to heat or direct sunlight should be avoided. Periodic inventories should be conducted, with unneeded items being discarded or returned to the storeroom/stockroom (225–6, 229).

3. Environmental Monitoring

Regular instrumental monitoring of airborne concentrations is not usually justified or practical in laboratories but may be appropriate when testing or redesigning hoods or other ventilation devices (12) or when a highly toxic substance is stored or used regularly (e.g., 3 times/week) (13).

4. Housekeeping, Maintenance, and Inspections

(a) *Cleaning.* Floors should be cleaned regularly (24).

(b) *Inspections.* Formal housekeeping and chemical hygiene inspections should be held at least quarterly (6, 21) for units which have frequent pesonnel changes and semiannually for others; informal inspections should be continual (21).

(c) *Maintenance.* Eye wash fountains should be inspected at intervals of not less than 3 months (6). Respirators for routine use should be inspected periodically by the laboratory supervisor (169). Safety showers should be tested routinely (169). Other safety equipment should be inspected regularly. (e.g., every 3–6 months) (6, 24, 171). Procedures to prevent restarting of out-of-service equipment should be established (25).

(d) *Passageways.* Stairways and hallways should not be used as storage areas (24). Access to exits, emergency equipment, and utility controls should never be blocked (24).

5. Medical Program

(a) *Compliance with regulations.* Regular medical surveillance should be established to the extent required by regulations (12).

(b) *Routine surveillance.* Anyone whose work involves regular and frequent handling of toxicologically significant quantities of a chemical should consult a qualified physician to determine on an individual basis whether a regular schedule of medical surveillance is desirable (11, 50).

(c) *First aid.* Personnel trained in first aid should be available during working hours and an emergency room with medical personnel should be nearby (173). See pp. 176–178 for description of some emergency first aid procedures.

6. Protective Apparel and Equipment

These should include for each laboratory:

(a) Protective apparel compatible with the required degree of protection for substances being handled (158–161);

(b) An easily accessible drench-type safety shower (162, 169);

(c) An eyewash fountain (162);

(d) A fire extinguisher (162–164);

(e) Respiratory protection (164–9), fire alarm and telephone for emergency use (162) should be available nearby; and

(f) Other items designated by the laboratory supervisor (156, 160).

7. Records

(a) Accident records should be written and retained (174).

(b) Chemical Hygiene Plan records should document that the facilities and precautions were compatible with current knowledge and regulations (7).

(c) Inventory and usage records for high-risk substances should be kept as specified in sections E3e below.

(d) Medical records should be retained by the institution in accordance with the requirements of state and federal regulations (12).

8. Signs and Labels

Prominent signs and labels of the following types should be posted:

(a) Emergency telephone numbers of emergency personnel/facilities, supervisors, and laboratory workers (28);

(b) Identity labels, showing contents of containers (including waste receptacles) and associated hazards (27, 48);

(c) Location signs for safety showers, eyewash stations, other safety and first aid equipment, exits (27) and areas where food and beverage consumption and storage are permitted (24); and

(d) Warnings at areas or equipment where special or unusual hazards exist (27).

9. Spills and Accidents

(a) A written emergency plan should be established and communicated to all personnel; it should include procedures for ventilation failure (200), evacuation, medical care, reporting, and drills (172).

(b) There should be an alarm system to alert people in all parts of the facility including isolation areas such as cold rooms (172).

(c) A spill control policy should be developed and should include consideration of prevention, containment, cleanup, and reporting (175).

(d) All accidents or near accidents should be carefully analyzed with the results distributed to all who might benefit (8, 28).

10. Information and Training Program

(a) Aim: To assure that all individuals at risk are adequately informed about the work in the laboratory, its risks, and what to do if an accident occurs (5, 15).

(b) Emergency and Personal Protection Training: Every laboratory worker should know the location and proper use of available protective apparel and equipment (154, 169).

Some of the full-time personnel of the laboratory should be trained in the proper use of emergency equipment and procedures (6).

Such training as well as first aid instruction should be available to (154) and encouraged for (178) everyone who might need it.

(c) Receiving and stockroom/storeroom personnel should know about hazards, handling equipment, protective apparel, and relevant regulations (217).

(d) Frequency of Training: The training and education program should be a regular, continuing activity—not simply an annual presentation (15).

(e) Literature/Consultation: Literature and consulting advice concerning chemical hygiene should be readily available to laboratory personnel, who should be encouraged to use these information resources (14).

11. Waste Disposal Program.

(a) Aim: To assure that minimal harm to people, other organisms, and the environment will result from the disposal of waste laboratory chemicals (5).

(b) Content (14, 232, 233, 240): The waste disposal program should specify how waste is to be collected, segregated, stored, and transported and include consideration of what materials may be incinerated. Transport from the institution must be in accordance with DOT regulations (244).

(c) Discarding Chemical Stocks: Unlabeled containers of chemicals and solutions should undergo prompt disposal; if partially used, they should not be opened (24, 27).

Before a worker's employment in the laboratory ends, chemicals for which that person was responsible should be discarded or returned to storage (226).

(d) Frequency of Disposal: Waste should be removed from laboratories to a central waste storage area at least once per week and from the central waste storage area at regular intervals (14).

(e) Method of Disposal: Incineration in an environmentally acceptable manner is the most practical disposal method for combustible laboratory waste (14, 238, 241).

Indiscriminate disposal by pouring waste chemicals down the drain (14, 231, 242) or adding them to mixed refuse for landfill burial is unacceptable (14).

Hoods should not be used as a means of disposal for volatile chemicals (40, 200).

Disposal by recycling (233, 243) or chemical decontamination (40, 230) should be used when possible.

E. Basic Rules and Procedures for Working with Chemicals

The Chemical Hygiene Plan should require that laboratory workers know and follow its rules and procedures. In addition to the procedures of the sub programs mentioned above, these should include the rules listed below.

1. General Rules

The following should be used for essentially all laboratory work with chemicals:

(a) Accidents and spills—Eye Contact: Promptly flush eyes with water for a prolonged period (15 minutes) and seek medical attention (33, 172).

Ingestion: Encourage the victim to drink large amounts of water (178).

Skin Contact: Promptly flush the affected area with water (33, 172, 178) and remove any contaminated clothing (172, 178). If symptoms persist after washing, seek medical attention (33).

Clean-up. Promptly clean up spills, using appropriate protective apparel and equipment and proper disposal (24 33). See pp. 233–237 for specific clean-up recommendations.

(b) Avoidance of "routine" exposure: Develop and encourage safe habits (23); avoid unnecessary exposure to chemicals by any route (23);

Do not smell or taste chemicals (32). Vent apparatus which may discharge toxic chemicals (vacuum pumps, distillation columns, etc.) into local exhaust devices (199).

Inspect gloves (157) and test glove boxes (208) before use.

Do not allow release of toxic substances in cold rooms and warm rooms, since these have contained recirculated atmospheres (209).

(c) Choice of chemicals: Use only those chemicals for which the quality of the available ventilation system is appropriate (13).

(d) Eating, smoking, etc.: Avoid eating, drinking, smoking, gum chewing, or application of cosmetics in areas where laboratory chemicals are present (22, 24, 32, 40); wash hands before conducting these activities (23, 24).

Avoid storage, handling or consumption of food or beverages in storage areas, refrigerators, glassware or utensils which are also used for laboratory operations (23, 24, 226).

(e) Equipment and glassware: Handle and store laboratory glassware with care to avoid damage; do not use damaged glassware (25). Use extra care with Dewar flasks and other evacuated glass apparatus; shield or wrap them to contain chemicals and fragments should implosion occur (25). Use equipment only for its designed purpose (23, 26).

(f) Exiting: Wash areas of exposed skin well before leaving the laboratory (23).

(g) Horseplay: Avoid practical jokes or other behavior which might confuse, startle or distract another worker (23).

(h) Mouth suction: Do not use mouth suction for pipeting or starting a siphon (23, 32).

(i) Personal apparel: Confine long hair and loose clothing (23, 158). Wear shoes at all times in the laboratory but do not wear sandals, perforated shoes, or sneakers (158).

(j) Personal housekeeping: Keep the work area clean and uncluttered, with chemicals and equipment being properly labeled and stored; clean up the work area on completion of an operation or at the end of each day (24).

(k) Personal protection: Assure that appropriate eye protection (154–158) is worn by all persons, including visitors, where chemicals are stored or handled (22, 23, 33, 154).

Wear appropriate gloves when the potential for contact with toxic materials exists (157); inspect the gloves before each use, wash them before removal, and replace them periodically (157). (A table of resistance to chemicals of common glove materials is given p. 159).

Use appropriate (164–168) respiratory equipment when air contaminant concentrations are not sufficiently restricted by engineering controls (164–5), inspecting the respirator before use (169).

Use any other protective and emergency apparel and equipment as appropriate (22, 157–162).

Avoid use of contact lenses in the laboratory unless necessary; if they are used, inform supervisor so special precautions can be taken (155).

Remove laboratory coats immediately on significant contamination (161).

(l) Planning: Seek information and advice about hazards (7), plan appropriate protective procedures, and plan positioning of equipment before beginning any new operation (22, 23).

(m) Unattended operations: Leave lights on, place an appropriate sign on the door, and provide for containment of toxic substances in the event of failure of a utility service (such as cooling water) to an unattended operation (27, 128).

(n) Use of hood: Use the hood for operations which might result in release of toxic chemical vapors or dust (198–9).

As a rule of thumb, use a hood or other local ventilation device when working with any appreciably volatile substance with a TLV of less than 50 ppm (13).

Confirm adequate hood performance before use: keep hood closed at all times except when adjustments within the hood are being made (200); keep materials stored in hoods to a minimum and do not allow them to block vents or air flow (200).

Leave the hood "on" when it is not in active use if toxic substances are stored in it or if it is uncertain whether adequate general laboratory ventilation will be maintained when it is "off" (200).

(o) Vigilance: Be alert to unsafe conditions and see that they are corrected when detected (22).

(p) Waste disposal: Assure that the plan for each laboratory operation includes plans and training for waste disposal (230).

Deposit chemical waste in appropriately labeled receptacles and follow all other waste disposal procedures of the Chemical Hygiene Plan (22, 24).

Do not discharge to the sewer concentrated acids or bases (231); highly toxic, malodorous, or lachrymatory substances

(231); or any substances which might interfere with the biological activity of waste water treatment plants, create fire or explosion hazards, cause structural damage or obstruct flow (242).

(q) *Working alone:* Avoid working alone in a building; do not work alone in a laboratory if the procedures being conducted are hazardous (28).

2. Working with Allergens and Embryotoxins

(a) *Allergens* (examples: diazomethane, isocyanates, bichromates): Wear suitable gloves to prevent hand contact with allergens or substances of unknown allergenic activity (35).

(b) *Embryotoxins* (34–5) (examples: organomercurials, lead compounds, formamide): If you are a woman of childbearing age, handle these substances only in a hood whose satisfactory performance has been confirmed, using appropriate protective apparel (especially gloves) to prevent skin contact.

Review each use of these materials with the research supervisor and review continuing uses annually or whenever a procedural change is made.

Store these substances, properly labeled, in an adequately ventilated area in an unbreakable secondary container.

Notify supervisors of all incidents of exposure or spills; consult a qualified physician when appropriate.

3. Work with Chemicals of Moderate Chronic or High Acute Toxicity

Examples: diisopropylflurophosphate (41), hydrofluoric acid (43), hydrogen cyanide (45).

Supplemental rules to be followed in addition to those mentioned above (Procedure B of "Prudent Practices", pp. 39–41):

(a) *Aim:* To minimize exposure to these toxic substances by any route using all reasonable precautions (39).

(b) *Applicability:* These precautions are appropriate for substances with moderate chronic or high acute toxicity used in significant quantities (39).

(c) *Location:* Use and store these substances only in areas of restricted access with special warning signs (40, 229).

Always use a hood (previously evaluated to confirm adequate performance with a face velocity of at least 60 linear feet per minute) (40) or other containment device for procedures which may result in the generation of aerosols or vapors containing the substance (39); trap released vapors to prevent their discharge with the hood exhaust (40).

(d) *Personal protection:* Always avoid skin contact by use of gloves and long sleeves (and other protective apparel as appropriate) (39). Always wash hands and arms immediately after working with these materials (40).

(e) *Records:* Maintain records of the amounts of these materials on hand, amounts used, and the names of the workers involved (40, 229).

(f) *Prevention of spills and accidents:* Be prepared for accidents and spills (41).

Assure that at least 2 people are present at all times if a compound in use is highly toxic or of unknown toxicity (39).

Store breakable containers of these substances in chemically resistant trays; also work and mount apparatus above such trays or cover work and storage surfaces with removable, absorbent, plastic backed paper (40).

If a major spill occurs outside the hood, evacuate the area; assure that cleanup personnel wear suitable protective apparel and equipment (41).

(g) *Waste:* Thoroughly decontaminate or incinerate contaminated clothing or shoes (41). If possible, chemically decontaminate by chemical conversion (40).

Store contaminated waste in closed, suitably labeled, impervious containers (for liquids, in glass or plastic bottles half-filled with vermiculite) (40).

4. Work with Chemicals of High Chronic Toxicity

(Examples: dimethylmercury and nickel carbonyl (48), benzo-a-pyrene (51), N-nitrosodiethylamine (54), other human carcinogens or substances with high carcinogenic potency in animals (38).)

Further supplemental rules to be followed, in addition to all these mentioned above, for work with substances of known high chronic toxicity (in quantities above a few milligrams to a few grams, depending on the substance) (47). (Procedure A of "Prudent Practices" pp. 47–50).

(a) *Access:* Conduct all transfers and work with these substances in a "controlled area": a restricted access hood, glove box, or portion of a lab, designated for use of highly toxic substances, for which all people with access are aware of the substances being used and necessary precautions (48).

(b) *Approvals:* Prepare a plan for use and disposal of these materials and obtain the approval of the laboratory supervisor (48).

(c) *Non-contamination/Decontamination:* Protect vacuum pumps against contamination by scrubbers or HEPA filters and vent them into the hood (49). Decontaminate vacuum pumps or other contaminated equipment, including glassware, in the hood before removing them from the controlled area (49, 50).

Decontaminate the controlled area before normal work is resumed there (50).

(d) *Exiting:* On leaving a controlled area, remove any protective apparel (placing it in an appropriate, labeled container) and thoroughly wash hands, forearms, face, and neck (49).

(e) *Housekeeping:* Use a wet mop or a vacuum cleaner equipped with a HEPA filter instead of dry sweeping if the toxic substance was a dry powder (50).

(f) *Medical surveillance:* If using toxicologically significant quantities of such a substance on a regular basis (e.g., 3 times per week), consult a qualified physician concerning desirability of regular medical surveillance (50).

(g) *Records:* Keep accurate records of the amounts of these substances stored (229) and used, the dates of use, and names of users (48).

(h) *Signs and labels:* Assure that the controlled area is conspicuously marked with warning and restricted access signs (49) and that all containers of these substances are

appropriately labeled with identity and warning labels (48).

(i) *Spills:* Assure that contingency plans, equipment, and materials to minimize exposures of people and property in case of accident are available (233–4).

(j) *Storage:* Store containers of these chemicals only in a ventilated, limited access (48, 227, 229) area in appropriately labeled, unbreakable, chemically resistant, secondary containers (48, 229).

(k) *Glove boxes:* For a negative pressure glove box, ventilation rate must be at least 2 volume changes/hour and pressure at least 0.5 inches of water (48). For a positive pressure glove box, thoroughly check for leaks before each use (49). In either case, trap the exit gases or filter them through a HEPA filter and then release them into the hood (49).

(l) *Waste:* Use chemical decontamination whenever possible; ensure that containers of contaminated waste (including washings from contaminated flasks) are transferred from the controlled area in a secondary container under the supervision of authorized personnel (49, 50, 233).

5. Animal Work with Chemicals of High Chronic Toxicity

(a) *Access:* For large scale studies, special facilities with restricted access are preferable (56).

(b) *Administration of the toxic substance:* When possible, administer the substance by injection or gavage instead of in the diet. If administration is in the diet, use a caging system under negative pressure or under laminar air flow directed toward HEPA filters (56).

(c) *Aerosol suppression:* Devise procedures which minimize formation and dispersal of contaminated aerosols, including those from food, urine, and feces (e.g., use HEPA filtered vacuum equipment for cleaning, moisten contaminated bedding before removal from the cage, mix diets in closed containers in a hood) (55, 56).

(d) *Personal protection:* When working in the animal room, wear plastic or rubber gloves, fully buttoned laboratory coat or jumpsuit and, if needed because of incomplete suppression of aerosols, other apparel and equipment (shoe and head coverings, respirator) (56).

(e) *Waste disposal:* Dispose of contaminated animal tissues and excreta by incineration if the available incinerator can convert the contaminant to non-toxic products (238); otherwise, package the waste appropriately for burial in an EPA-approved site (239).

F. Safety Recommendations

The above recommendations from "Prudent Practices" do not include those which are directed primarily toward prevention of physical injury rather than toxic exposure. However, failure of precautions against injury will often have the secondary effect of causing toxic exposures. Therefore, we list below page references for recommendations concerning some of the major categories of safety hazards which also have implications for chemical hygiene:

1. Corrosive agents: (35–6)

Federal Register / Vol. 55, No. 21 / Wednesday, January 31, 1990 / Rules and Regulations   3335

2. Electrically powered laboratory apparatus: (179–92)

3. Fires, explosions: (26, 57–74, 162–4, 174–5, 219–20, 226–7)

4. Low temperature procedures: (26, 88)

5. Pressurized and vacuum operations (including use of compressed gas cylinders): (27, 75–101)

G. Material Safety Data Sheets

Material safety data sheets are presented in "Prudent Practices" for the chemicals listed below. (Asterisks denote that comprehensive material safety data sheets are provided).

*Acetyl peroxide (105)
*Acrolein (106)
*Acrylonitrile (107)
Ammonia (anhydrous) (91)
*Aniline (109)
*Benzene (110)
*Benzo[a]pyrene (112)
*Bis(chloromethyl) ether (113)
Boron trichloride (91)
Boron trifluoride (92)
Bromine (114)
*Tert-butyl hydroperoxide (148)
*Carbon disulfide (116)
Carbon monoxide (92)
*Carbon tetrachloride (118)
*Chlorine (119)
Chlorine trifluoride (94)
*Chloroform (121)
Chloromethane (93)
*Diethyl ether (122)
Diisopropyl fluorophosphate (41)
*Dimethylformamide (123)
*Dimethyl sulfate (125)
*Dioxane (126)
*Ethylene dibromide (128)
*Fluorine (95)
*Formaldehyde (130)
*Hydrazine and salts (132)
Hydrofluoric acid (43)
Hydrogen bromide (98)
Hydrogen chloride (98)
*Hydrogen cyanide (133)
*Hydrogen sulfide (135)
Mercury and compounds (52)
*Methanol (137)
*Morpholine (138)
*Nickel carbonyl (99)
*Nitrobenzene (139)
Nitrogen dioxide (100)
N-nitrosodiethylamine (54)
*Peracetic acid (141)
*Phenol (142)
*Phosgene (143)
*Pyridine (144)
*Sodium azide (145)
*Sodium cyanide (147)
Sulfur dioxide (101)
*Trichloroethylene (149)
*Vinyl chloride (150)

Appendix B to § 1910.1450—References (Non-Mandatory)

The following references are provided to assist the employer in the development of a Chemical Hygiene Plan. The materials listed below are offered as non-mandatory guidance. References listed here do not imply specific endorsement of a book, opinion, technique, policy or a specific solution for a safety or health problem. Other references not listed here may better meet the needs of a specific laboratory. (a) Materials for the development of the Chemical Hygiene Plan:

1. American Chemical Society. Safety in Academic Chemistry Laboratories, 4th edition. 1985.

2. Fawcett, H.H. and W. S. Wood, Safety and Accident Prevention in Chemical Operations. 2nd edition. Wiley-Interscience. New York, 1982.

3. Flury, Patricia A., Environmental Health and Safety in the Hospital Laboratory, Charles C. Thomas Publisher, Springfield IL, 1978.

3. Green, Michael E. and Turk, Amos. Safety in Working with Chemicals. Macmillan Publishing Co., NY, 1978.

5. Kaufman, James A., Laboratory Safety Guidelines, Dow Chemical Co., Box 1713, Midland, MI 48640, 1977.

6. National Institutes of Health, NIH Guidelines for the Laboratory use of Chemical Carcinogens, NIH Pub. No. 81–2385, GPO, Washington, DC 20402, 1981.

7. National Research Council. Prudent Practices for Disposal of Chemicals from Laboratories, National Academy Press, Washington, DC, 1983.

8. National Research Council, Prudent Practices for Handling Hazardous Chemicals in Laboratories, National Academy Press, Washington, DC, 1981.

9. Renfrew, Malcolm, Ed., Safety in the Chemical Laboratory, Vol. IV, J. Chem. Ed., American Chemical Society, Easion, PA, 1981.

10. Steere, Norman V., Ed., Safety in the Chemical Laboratory, J. Chem. Ed. American Chemical Society, Easion, PA, 18042, Vol. I, 1967, Vol. II, 1971, Vol. III 1974.

11. Steere, Norman V., Handbook of Laboratory Safety, the Chemical Rubber Company Cleveland, OH, 1971.

12. Young, Jay A., Ed., Improving Safety in the Chemical Laboratory, John Wiley & Sons, Inc. New York, 1987.

(b) Hazardous Substances Information:

1. American Conference of Governmental Industrial Hygienists, Threshold Limit Values for Chemical Substances and Physical Agents in the Workroom Environment with Intended Changes, P.O. Box 1937 Cincinnati, OH 45201 (latest edition).

2. Annual Report on Carcinogens, National Toxicology Program U.S. Department of Health and Human Services, Public Health Service, U.S. Government Printing Office, Washington, DC, (latest edition).

3. Best Company, Best Safety Directory, Vols. I and II, Oldwick, N.J., 1981.

4. Bretherick, L., Handbook of Reactive Chemical Hazards. 2nd edition. Butterworths. London, 1979.

5. Bretherick, L., Hazards in the Chemical Laboratory, 3rd edition. Royal Society of Chemistry, London. 1986.

6. Code of Federal Regulations. 29 CFR part 1910 subpart Z. U.S. Govt. Printing Office. Washington, DC 20402 (latest edition).

7. IARC Monographs on the Evaluation of the Carcinogenic Risk of Chemicals to Man, World Health Organization Publications Center, 49 Sheridan Avenue, Albany, New York 12210 (latest editions).

8. NIOSH/OSHA Pocket Guide to Chemical Hazards. NIOSH Pub. No. 85–114, U.S. Government Printing Office. Washington, DC, 1985 (or latest edition).

9. Occupational Health Guidelines, NIOSH/OSHA NIOSH Pub. No. 81–123 U.S. Government Printing Office. Washington, DC. 1981.

10. Patty, F.A., Industrial Hygiene and Toxicology, John Wiley & Sons, Inc., New York, NY (Five Volumes).

11. Registry of Toxic Effects of Chemical Substances, U.S. Department of Health and Human Services, Public Health Service, Centers for Disease Control, National Institute for Occupational Safety and Health, Revised Annually, for sale from Superintendent of Documents U.S. Govt. Printing Office. Washington, DC 20402.

12. The Merck Index: An Encyclopedia of Chemicals and Drugs. Merck and Company Inc. Rahway, N.J., 1976 (or latest edition).

13. Sax, N.I. Dangerous Properties of Industrial Materials, 5th edition, Van Nostrand Reinhold, NY., 1979.

14. Sittig, Marshall, Handbook of Toxic and Hazardous Chemicals, Noyes Publications, Park Ridge, NJ, 1981.

(c) Information on Ventilation:

1. American Conference of Governmental Industrial Hygienists Industrial Ventilation, 16th edition Lansing, MI. 1980.

2. American National Standards Institute, Inc. American National Standards Fundamentals Governing the Design and Operation of Local Exhaust Systems ANSI Z 9.2–1979 American National Standards Institute, N.Y. 1979.

3. Imad, A.P. and Watson, C.L. Ventilation Index: An Easy Way to Decide about Hazardous Liquids, Professional Safety pp 15–18, April 1980.

4. National Fire Protection Association. Fire Protection for Laboratories Using Chemicals NFPA–45, 1982.

Safety Standard for Laboratories in Health Related Institutions. NFPA, 56c. 1980.

Fire Protection Guide on Hazardous Materials. 7th edition. 1978.

National Fire Protection Association. Batterymarch Park, Quincy, MA 02269.

5. Scientific Apparatus Makers Association (SAMA), Standard for Laboratory Fume Hoods, SAMA LF7–1980, 1101 16th Street, NW., Washington, DC 20036.

(d) Information on Availability of Referenced Material:

1. American National Standards Institute (ANSI), 1430 Broadway, New York, NY 10018.

2. American Society for Testing and Materials (ASTM), 1916 Race Street, Philadelphia, PA 19103.

(Approved by the Office of Management and Budget under control number 1218–0131)

[FR Doc. 90–1717 Filed 1–30–90: 8:45 am]

BILLING CODE 4510-26-M

APPENDIX C

EPA Levels of Protection

Level A protection should be worn when the highest level of respiratory, skin, eye, and mucous membrane protection is needed.

Level A Conditions:
Unknown gaseous concentrations.
Known gaseous contaminants in high concentrations.
Known gaseous contaminants that are extremely toxic.
Possible or expected skin exposure to toxic substances.

Recommended Protective Equipment:
Positive-pressure self-contained breathing apparatus (MSHA/NIOSH approved).
Fully encapsulating chemical-resistant suit.
Gloves, inner, chemical-resistant.
Gloves, outer, chemical-resistant.
Boots, chemical-resistant, steel toe and shank (depending on suit boot construction, worn over or under suit boot).
Underwear, cotton, longjohn type.
Two-way radio communication (intrinsically safe).

Optional Protective Equipment:
Hard hats (under suit)
Coveralls (under suit)

Level B protection should be selected when the highest level of respiratory protection is needed, but a lesser level of skin and eye protection. Level B protection is the minimum level recommended on initial site entries until the hazards have been further identified by monitoring, sampling, and other reliable methods of analysis, and personal protective equipment corresponding with those findings has been utilized.

Level B Conditions:
Known contaminant levels, and these contamination levels exceed the limit of air-purifying devices.

Atmosphere with less than 19.5% (by volume) of oxygen.
Atmosphere with chemical concentrations above the IDLH level.

Recommended Protective Equipment:
Positive-pressure self-contained breathing apparatus (MSHA/NIOSH approved).
Chemical-resistant clothing (overalls and long-sleeved jacket, cover alls, hooded two piece chemical splash suit, disposable chemical-resistant coveralls.
Gloves, inner, chemical-resistant.
Gloves, outer, chemical-resistant.
Boots, steel toe and shank, chemical-resistant.
Two-way radio communications (intrinsically safe).

Optional Protective Equipment:
Hard hats.
Coveralls (under splash suit).
Boots, outer, chemical-resistant.

Level C protection should be selected when the type of airborne substance is known, concentration measured, criteria for using air-purifying respirators met, and skin and eye exposure is unlikely. Periodic monitoring of the air must be performed.

Level C Conditions:
Atmosphere with greater than 19.5% (by volume) of oxygen.
Atmosphere with chemical contaminant levels below IDLH level.
Chemical contaminants have adequate warning properties.
Chemical contaminants to unprotected body areas such as head and neck are within skin exposure guidelines.
Skin contact hazards are minimal or do not exist.

Recommended Protective Equipment:
Full-face, air-purifying respirator (MSHA/NIOSH approved).
Chemical-resistant clothing (one-piece coverall, hooded two-piece chemical splash suit, chemical resistant hood and apron, disposable chemical-resistant coveralls.
Gloves, outer, chemical-resistant.
Boots, steel toe and shank, chemical-resistant.
Two-way radio communications (intrinsically safe).

Optional Protective Equipment:
Gloves, inner, chemical-resistant.
Boots, outer, chemical-resistant.

Cloth coveralls (inside chemical protective clothing).
Hard hats.
Escape mask.

Level D protection is primarily a work uniform. It should not be worn on any site where respiratory or skin hazards exist.

Level D Conditions:
No possibility for respiratory exposure.
No possibility for skin contamination.

Recommended Protective Equipment:
Coveralls
Boots/shoes, safety or chemical-resistant, steel toe and shank.

Optional Protective Equipment:
Safety glasses or chemical splash goggles.
Gloves.
Boots, outer, chemical-resistant, disposable.
Hard hats with face shield.
Escape mask.

APPENDIX D

Vapor Suppression Foams

National Foam Company markets three foam concentrates that have been tested to be effective on the following hazardous chemicals. The lists do not include those chemicals that are yet to be tested. Contact the company for the latest information.

HAZMAT NF® No. 1 has been tested effective on the following vapors:
Alkyl amines
Ammonia solutions
Anhydrous ammonia
Dimethyl amine
Ethyl amines
Hydrazine
Methyl amines
Sodium hypochlorite
Triethylamine
Trimethylamine

HAZMAT NF® No. 2 has been tested effective on the following vapors:
Bromine
Chlorine
Dimethyldichlorosilane
Hydrochloric acid (all strengths)
Hydrofluoric acid (70%)
Hydrogen chloride (anhydrous)
Methyl chloroacetate
Nitric acid (all strengths)
Sulfur dioxide
Sulfur monochloride
Titanium tetrachloride
Trichlorosilane

Universal® has been tested effective for the following vapors:

Acetaldehyde
Acetaldoxime
Acetic acid
Acetic anhydride
Acetone
Acetone cyanohydrin
Acrolein
Acrylonitrile
Aldehydes
Amines
Benzene
Butanediol
Butynediol
Chlorinated hydrocarbons
Cyclohexyl isocyanate
Cyclohexane
Dichloroaniline
Dimethyl formamide
Epichlorohydrin
Esters
Ethers
Ethyl acetate
Ethyl benzene
Ethylenediamine
Ethylene oxide
Hydrazine monoacetate
Hydrocarbons
Hydrogen peroxide
Hydroquinone in methanol
Hydroquinone in vinyl acetate
Ketones
Mercaptans
Methyl acetate
Methyl chloroacetate
Methylene chloride
Methyl isocyanate
Methyl methacrylate
Monochlorobenzene
Monomers
Propylene oxide
Styrene
Tetrahydrofuran
Tars
Toluene
Vinyl acetate

Appendix E

EPA-Listed RCRA Hazardous Wastes

Waste Number	Hazardous Waste	Hazard Code

Hazardous Waste From Nonspecific Sources

F001   The following spent halogenated solvents used in degreasing: (T)
tetrachloroethylene, trichloroethylene, methylene chloride,
1,1,1-trichloroethane, carbon tetrachloride, chlorinated
fluorocarbons, all spent solvent mixture/blends used in
degreasing containing, before use, a total of ten percent or
more (by volume) of one or more of the above halogenated
solvents or those solvents listed in F002, F004, and F005; and
still bottoms from the recovery of these spent solvents and
spent solvent mixtures.

F002   The following spent halogenated solvents: tetrachloroethylene,   (T)
methylene chloride, trichloroethylene, 1,1,1-trichloroethane,
chlorobenzene, 1,1,2-trichloro-1.2,2-trifluoroethane, o-
dichlorobenzene, and trichlorofluoromethane; all spent solvent
mixtures/blends containing, before use, a total of ten percent
or more of the above halogenated solvents or those listed in
F001, F004, or F005; and still bottoms from the recovery of
these spent solvents and spent solvent mixtures.

F003   The following spent nonhalogenated solvents: xylene, acetone,   (I)
ethyl acetate, ethyl benzene, ethyl ether, methyl isobutyl
ketone, n-butyl alcohol, cyclohexanone, methanol; all spent
solvent mixtures/blends containing, before use, one or more of
the above nonhalogenated solvents, and, a total of ten percent
or more (by volume) of one or more of those solvents listed in
F001, F002, F004, and F005; and still bottoms from the
recovery of these spent solvents and spent solvent mixtures.

F004   The following spent nonhalogenated solvents: cresols and   (T)
cresylic acid, nitrobenzene; all spent solvent mixtures/blends
containing, before use, a total of ten percent or more (by

Hazard Codes: (C) Corrosive; (H) Acutely Hazardous; (I) Ignitable; (R) Reactive; and (T) Toxic.

Waste Number	Hazardous Waste	Hazard Code
	volume) of one or more of the above nonhalogenated solvents or those solvents listed in F001, F002, and F005; and the still bottoms from the recovery of these spent solvents and spent solvent mixtures.	
F005	The following spent nonhalogenated solvents: toluene, methyl ethyl ketone, carbon disulfide, isobutanol, pyridine; all spent solvent mixtures/blends containing, before use, a total of ten percent or more (by volume) of one or more of the above nonhalogenated solvents or those listed in F001, F002, and F004; and the still bottoms from the recovery of these spent solvents and spent solvent mixtures.	(I,T)
F006	Wastewater treatment sludges from electroplating operations except from the following processes: (1) sulfuric acid anodizing of aluminum; (2) tin plating on carbon steel; (3)zinc plating (segregated basis) on carbon steel; (4) aluminum or zinc-aluminum plating on carbon steel; (5) cleaning/stripping associated with tin, zinc, and aluminum plating on carbon steel; and (6) chemical etching and milling of aluminum	(T)
F019	Wastewater treatment sludges from the chemical conversion coating of aluminum	(T)
F007	Spent cyanide plating bath solutions from electroplating operations (except for precious metals electroplating spent cyanide plating bath solutions)	(R,T)
F008	Plating bath sludges from the bottom of plating baths from electroplating operations for which cyanides are used in the process (except for precious metals electroplating plating bath sludges)	(R,T)
F009	Spent stripping and cleaning bath solutions from electroplating operations for which cyanides are used in the process (except for precious metals electroplating spent stripping and cleaning bath solutions)	(R,T)
F010	Quenching bath sludges from oil baths from metal heat treating operations for which cyanides are used in the process (except for precious metals heat-treating quenching bath sludges)	(R,T)
F011	Spent cyanide solutions from salt bath pot cleaning from metal heat treating operations (except for precious metals heat treating spent cyanide solutions from salt bath pot cleaning)	(R,T)
F012	Quenching wastewater treatment sludges from metal heat treating operations for which cyanides are used in the process	(T)

Waste Number	Hazardous Waste	Hazard Code
	(except for precious metals heat treating quenching wastewater treatment sludges)	
F024	Wastes including but not limited to distillation residues, heavy ends, tars, and reactor clean-out wastes from the production of chlorinated aliphatic hydrocarbons, having carbon content from one to five, utilizing free radical catalyzed processes (Does not include light ends, spent filters and filter aids, spent dessicants, wastewater, wastewater treatment sludges, spent catalysts and wastes listed in 261.32)	(T)
F020	Wastes (except wastewater and spent carbon from hydrogen chloride purification) from the production or manufacturing use (as a reactant, chemical intermediate, or component in a formulating process) of tri- or tetrachlorophenol or of intermediates used to produce their pesticide derivatives (Does not include wastes from the production of Hexachlorophene from highly purified 2,4,5-trichlorophenol)	(H)
F021	Wastes (except wastewater and spent carbon from hydrogen chloride purification) from the production or manufacturing use (as a reactant, chemical intermediate, or component in a formulating process) of pentachlorophenol or of intermediates used to produce its derivatives	(H)
F022	Wastes (except wastewater and spent carbon from hydrogen chloride purification) from the manufacturing use (as a reactant, chemical intermediate, or component in a formulating process) of tetra-, penta-, or hexachlorobenzenes under alkaline conditions	(H)
F023	Wastes (except wastewater and spent carbon from hydrogen chloride purification) from the production of materials on equipment previously used for the production or manufacturing use (as a reactant, chemical intermediate, or component in a formulating process) of tri- and tetrachlorophenols (Does not include wastes from equipment used only for the production or use of hexachlorophene from highly purified 2,4,5-trichlorophenol)	(H)
F026	Wastes (except wastewater and spent carbon from hydrogen chloride purification) from the production of materials on equipment previously used for the manufacturing use (as a reactant, chemical intermediate, or component in a formulating process) of tetra-, penta-, or hexachlorobenzene under alkaline conditions	(H)
F027	Discarded unused formulations containing tri-, tetra-, or	(H)

Waste Number	Hazardous Waste	Hazard Code

pentachlorophenol or discarded unused formulations
containing compounds derived from these chlorophenols
(Does not include formulations containing hexachlorophene
synthesized from prepurified 2,4,5-trichlorophenol as the sole
component

F028   Residues resulting from the incineration or thermal treatment   (T)
        of soil contaminated with EPA hazardous wastes numbered
        F020, F021, F022, F023, F026, and F027

Hazardous Wastes From Specific Sources

Wood Preservatives

K001   Bottom sediment sludge from the treatment of wastewaters   (T)
        from wood preserving processes that use creosote and/or
        pentachlorophenol

Inorganic Pigments

K002   Wastewater treatment sludge from the production of chrome   (T)
        yellow and orange pigments

K003   Wastewater treatment sludge from the production of   (T)
        molybdate orange pigments

K004   Wastewater treatment sludge from the production of zinc   (T)
        yellow pigments

K005   Wastewater treatment sludge from the production of chrome   (T)
        green pigments

K006   Wastewater treatment sludge from the production of chrome   (T)
        oxide green pigments (anhydrous and hydrated)

K007   Wastewater treatment sludge from the production of iron blue   (T)
        pigments

K008   Oven residue from the production of chrome oxide green   (T)
        pigments

Waste Number	Hazardous Waste	Hazard Code

Organic Chemicals

K009	Distillation bottoms from the production of acetaldehyde from ethylene	(T)
K010	Distillation side cuts from the production of acetaldehyde from ethylene	(T)
K011	Bottom stream from the wastewater stripper in the production of acrylonitrile	(H)
K013	Bottom stream from the acetonitrile column in the production of acrylonitrile	(R,T)
K014	Bottoms from the acetonitrile purification column in the production of acrylonitrile	(T)
K015	Still bottoms from the distillation of benzyl chloride	(T)
K016	Heavy ends or distillation residues from the production of carbon tetrachloride	(T)
K017	Heavy ends (still bottoms) from the purification column in the production of epichlorohydrin	(T)
K018	Heavy ends from the fractionation column in ethyl chloride production	(T)
K019	Heavy ends from the distillation of ethylene dichloride in ethylene dichloride production	(T)
K020	Heavy ends from the distillation of vinyl chloride in vinyl chloride monomer production	(T)
K021	Aqueous spent antimony catalyst waste from fluoromethanes production	(T)
K022	Distillation bottom tars from the production of phenol/ acetone from cumene	(T)
K023	Distillation light ends from the production of phthalic anhydride from naphthalene	(T)
K024	Distillation bottoms from the production of phthalic anhydride from naphthalene	(T)
K093	Distillation light ends from the production of phthalic anhydride from o-xylene	(T)
K094	Distillation bottoms from the production of phthalic anhydride from o-xylene	(T)

Waste Number	Hazardous Waste	Hazard Code
K025	Distillation bottoms from the production of nitrobenzene by the nitration of benzene	(T)
K026	Stripping still tails from the production of methyl ethyl pyridines	(T)
K027	Centrifuge and distillation residues from toluene diisocyanate production	(R,T)
K028	Spent catalyst from the hydrochlorinator reactor in the production of 1,1,1-trichloroethane	(T)
K029	Waste from the product steam stripper in the production of 1,1,1-trichloroethane	(T)
K095	Distillation bottoms from the production of 1,1,1-trichloroethane	(T)
K096	Heavy ends from the heavy ends column from the production of 1,1,1-trichloroethane	(T)
K030	Column bottoms or heavy ends from the combined production of trichloroethylene and perchloroethylene	(T)
K083	Distillation bottoms from aniline production	(T)
K103	Process residues from aniline extraction from the production of aniline	(T)
K104	Combined wastewater streams generated from nitrobenzene/aniline production	(T)
K085	Distillation or fractionation column bottoms from the production of chlorobenzenes	(T)
K105	Separated aqueous stream from the reactor product washing step in the production of chlorobenzenes	(T)
K111	Product washwaters from the production of dinitrotoluene via nitration of toluene	(C,T)
K112	Reaction by-product water from the drying column in the production of toluenediamine via hydrogenation of dinitrotoluene	(T)
K113	Condensed liquid light ends from the purification of toluenediamine in the production of toluenediamine via hydrogenation of dinitrotoluene	(T)
K114	Vicinals from the purification of toluenediamine in the production of toluenediamine via hydrogenation of dinitrotoluene	(T)

Waste Number	Hazardous Waste	Hazard Code
K115	Heavy ends from the purification of toluenediamine in the production of toluenediamine via hydrogenation of dinitrotoluene	(T)
K116	Organic condensate from the solvent recovery column in the production of toluene diisocyanate via phosgenation of toluenediamine	(T)
K117	Wastewater from the reactor vent gas scrubber in the production of ethylene dibromide via bromination of ethene	(T)
K118	Spent adsorbent solids from purification of ethylene dibromide via bromination of ethene	(T)
K136	Still bottoms from the purification of ethylene dibromide in the production of ethylene dibromide via bromination of ethene	(T)

Inorganic Chemicals

K071	Brine purification muds from the mercury cell process in chlorine production for which separately prepurified brine is not used	(T)
K073	Chlorinated hydrocarbon waste from the purification step of the diaphragm cell process using graphite anodes in chlorine production	(T)
K106	Wastewater treatment sludge from the mercury cell process in chlorine production	(T)

Pesticides

K031	By-product salts generated in the production of MSMA and cacodylic acid	(T)
K032	Wastewater treatment sludge from the production of chlordane	(T)
K033	Wastewater and scrub water from the chlorination of cyclo-pentadiene in the production of chlordane	(T)
K034	Filter solids from the filtration of hexachlorocyclopentadiene in the production of chlordane	(T)
K097	Vacuum stripper discharge from the chlordane chlorinator in the production of chlordane	(T)

Waste Number	Hazardous Waste	Hazard Code
K035	Wastewater treatment sludges generated in the production of creosote	(T)
K036	Still bottoms from toluene reclamation distillation in the production of disulfoton	(T)
K037	Wastewater treatment sludges from the production of disulfoton	(T)
K038	Wastewater from the washing and stripping of phorate production	(T)
K039	Filter cake from the distillation of diethylphosphorodithioic acid in the production of phorate	(T)
K040	Wastewater treatment sludge from the production of phorate	(T)
K041	Wastewater treatment sludge from the production of toxaphene	(T)
K098	Untreated process wastewater from the production of toxaphene	(T)
K042	Heavy ends or distillation residues from the distillation of tetrachlorobenzene in the production of 2,4,5-T	(T)
K043	2,6-Dichlorophenol waste from the production of 2,4-D	(T)
K099	Untreated wastewater from the production of 2,4-D	(T)

Explosives

K044	Wastewater treatment sludges from the manufacturing and processing of explosives	(R)
K045	Spent carbon from the treatment of wastewater containing explosives	(R)
K046	Wastewater treatment sludges from the manufacturing, formulation, and loading of lead-based initiating compounds	(R)
K047	Pink/red water from TNT operations	(R)

Petroleum Refining

K048	Dissolved air floatation (DAF) float from the petroleum refining industry	(T)
K049	Slop oil emulsion solids from the petroleum refining industry	(T)

Waste Number	Hazardous Waste	Hazard Code
K050	Heat exchanger bundle cleaning sludge from the petroleum refining industry	(T)
K051	API separator sludge from the petroleum refining industry	(T)
K052	Tank bottoms (leaded) from the petroleum refining industry	(T)

Iron and Steel

K061	Emission control dust/sludge from the primary production of steel in electric furnaces	(T)
K062	Spent pickle liquor generated by steel finishing operations of facilities within iron and steel industry SIC codes 331 and 332.	(C,T)

Secondary Lead

K069	Emission control dust/sludge from secondary lead smelting	(T)
K100	Waste leaching solution from acid leaching of emission control dust/sludge from secondary lead smelting	(T)

Veterinary Pharmaceuticals

K084	Wastewater treatment sludges generated during the production of veterinary pharmaceuticals from arsenic or organo-arsenic compounds	(T)
K101	Distillation tar residues from the distillation of aniline-based compounds in the production of veterinary pharmaceuticals from arsenic or organo-arsenic compounds	(T)
K102	Residue from the use of activated carbon for decolorization in the production of veterinary pharmaceuticals from arsenic or organo-arsenic compounds	(T)

Ink Formulation

K086	Solvent washes and sludges, caustic washes and sludges, or water washes and sludges from cleaning tubs and equipment	(T)

Waste Number	Hazardous Waste	Hazard Code

used in the formulation of ink from pigments, driers, soaps, and stabilizers containing chromium and lead

Coking

K060	Ammonia still lime sludge from coking operations	(T)
K087	Decanter tank tar sludge from coking operations	(T)

Commercial Chemical Products

The following P code wastes are considered acutely hazardous.

P023   Acetaldehyde, chloro-
P002   Acetamide, N-(aminothioxomethyl)-
P057   Acetamide, 2-fluoro-
P058   Acetic acid, fluoro-, sodium salt
P066   Acetimidic acid, N-[(methylcarbamoyl)oxy]thio-, methyl ester
P001   3-(alpha-acetonylbenzyl)-4-hydroxycoumarin and salts, when present at
      concentrations greater than 0.3%
P002   1-Acetyl-2-thiourea
P003   Acrolein
P070   Aldicarb
P004   Aldrin
P005   Allyl alcohol
P006   Aluminum phosphide
P007   5-(Aminomethyl)-3-isoxazolol
P008   4-aAminopyridine
P009   Ammonium picrate (R)
P119   Ammonium vanadate
P010   Arsenic acid
P012   Arsenic(III) oxide
P011   Arsenic (V) oxide
P011   Arsenic pentoxide
P012   Arsenic trioxide
P038   Arsine, diethyl
P054   Aziridine

P013   Barium cyanide
P024   Benzenamine, 4-chloro-
P077   Benzenamine, 4-nitro-
P028   Benzene, (chloromethyl)-
P042   1,2-Benzenediol, 4-[(1-hydroxy-2-(methyl-amino)ethyl)]-
P014   Benzenethiol
P028   Benzyl chloride
P015   Beryllium dust
P016   Bis(chloromethyl) ether
P017   Bromoacetone
P018   Brucine

P021   Calcium cyanide
P123   Camphene, octachloro-
P103   Carbamimidoselenoic acid
P022   Carbon bisulfide
P022   Carbon disulfide

P095 Carbonyl chloride
P033 Chlorine cyanide
P023 Chloroacetaldehyde
P024 p-Chloroaniline
P026 1-(o-Chlorophenyl)thiourea
P027 3-Chloropropionitrile
P029 Copper cyanides
P030 Cyanides (soluble cyanide salts), not elsewhere specified
P031 Cyanogen
P033 Cyanogen chloride

P036 Dichlorophenylarsine
P037 Dieldrin
P038 Diethylarsine
P039 O,O-Diethyl S-[2-(ethylthio)ethyl] phosphorodithioate
P041 Diethyl-p-nitrophenyl phosphate
P040 O,O-Diethyl O-pyrazinyl phosphorothioate
P043 Diisopropyl fluorophosphate
P044 Dimethoate
P045 3,3-Dimethyl-1-(methylthio)-2-butanone,O-[(methylamino)carbonyl]
      oxime
P071 O,O-Dimethyl O-p-nitrophenyl phosphorothioate
P082 Dimethylnitrosamine
P046 alpha,alpha-Dimethylphenethylamine
P047 4,6-Dinitro-o-cresol and salts
P034 4,6-Dinitro-o-cyclohexylphenol
P048 2,4-Dinitrophenol
P020 Dinoseb
P085 Diphosphoramide, octamethyl
P039 Disulfoton
P049 2,4-Dithiobiuret
P109 Dithiopyrophosphoric acid, tetraethyl ester

P050 Endosulfan
P088 Endothall
P051 Endrin
P042 Epinephrine
P046 Ethanamine, 1,1-dimethyl-2-phenyl-
P084 Ethenamine, N-methyl-N-nitroso-
P101 Ethyl cyanide
P054 Ethylenimine

P097 Famphur
P056 Fluorine
P057 Fluoroacetamide
P058 Fluoroacetic acid, sodium salt
P065 Fulminic acid, mercury(II) salt

P059 Heptachlor
P051 1,2,3,4,10,10-Hexachloro-6,7-epoxy-1,4,4a,5,6,7,8,8a-octahydro-endo,endo-1,4:5,8-dimethanonaphthalene
P037 1,2,3,4,10,10,-Hexachloro-6,7-epoxy-1,4,4a,5,6,7,8,8a-octahydro-endo, exo-1,4:5,8-dimethanonaphthalene
P060 1,2,3,4,10,10-Hexachloro-1,4,4a,5,8,8a-hexahydro-1,4:5,8-endo, endo-dimethanonaphthalene
P004 1,2,3,4,10,10-Hexachloro-1,4,4a,5,8,8a-hexahydro-1,4:5,8-endo, exo-dimethanonaphthalene
P060 Hexachlorohexahydro-exo, exo-dimethanonaphthalene P062 Hexaethyl tetraphosphate
P116 Hydrazinecarbothioamide
P068 Hydrazine, methyl-
P063 Hydrocyanic acid
P063 Hydrogen cyanide
P096 Hydrogen phosphide

P064 Isocyanic acid, methyl ester
P007 3(2H)-isoxazolone, 5-(aminomethyl)-

P092 Mercury, (acetato-O)phenyl-
P065 Mercury fulminate (R,T)
P016 Methane, oxybis(chloro)-
P112 Methane, tetranitro- (R)
P118 Methanethiol, trichloro-
P059 4,7-Methano-1H-indene, 1,4,5,6,7,8,8-heptachloro-3a,4,7,7a-tetrahy-dro-
P066 Methomyl
P067 2-Methylaziridine
P068 Methyl hydrazine
P064 Methyl isocyanate
P069 2-Methyllactonitrile
P071 Methyl parathion

P072 alpha-Naphthylthiourea
P073 Nickel carbonyl
P074 Nickle cyanide
P074 Nickle(II) cyanide
P073 Nickle tetracarbonyl
P075 Nicotine and salts
P076 Nitric oxide
P077 p-Nitroaniline
P078 Nitrogen dioxide
P076 Nitrogen(II) oxide
P078 Nitrogen(IV) oxide
P081 Nitroglycerine (R)
P082 N-Nitrosodimethylamine

P084 N-Nitrosomethylvinylamine
P050 5-Norbornene-2,3-dimethanol,1,4,5,6,7,7-hexachloro, cyclic sulfite

P085 Octamethylpyrophosphoramide
P087 Osmium oxide
P087 Osmium tetroxide
P088 7-Oxabicyclo-[2.2.1] heptane-2,3-dicarboxylic acid

P089 Parathion
P034 Phenol, 2-cyclohexyl-4,6-dinitro-
P048 Phenol, 2,4-dinitro-
P047 Phenol, 2,4-dinitro-6-methyl-
P020 Phenol, 2,4-dinitro-6-(1-methylpropyl)-
P009 Phenol, 2,4,6-trinitro-, ammonium salt (R)
P036 Phenyl dichloroarsine
P092 Phenylmercuric acetate
P093 N-Phenylthiourea
P094 Phorate
P095 Phosgene
P096 Phosphine
P041 Phosphoric acid, diethyl p-nitrophenyl ester
P044 Phosphorodithioic acid, O,O-dimethyl S-[2-(methylamino)-2-oxyethyl]
         ester
P043 Phosphorofluoric acid, bis(1-methylethyl)ester
P094 Phosphorothioic acid, O,O-diethyl S-(ethylthio) methyl ester
P089 Phosphorothioic acid, O,O-diethyl O(p-nitrophenyl) ester
P040 Phosphorothioic acid, O,O-diethyl O-pyrazinyl ester
P097 Phosphorothioic acid, O,O-dimethyl O-[p(dimethylamino)-sulfonyl)
         phenyl] ester
P110 Plumbane, tetraethyl-
P098 Potassium cyanide
P099 Potassium silver cyanide
P070 Propanal, 2-methyl-2-(methylthio)-O-[(methylamino) carbonyl]oxime
P101 Propanenitrile
P027 Propanenitrile, 3-chloro
P069 Propanenitrile, 2-hydroxy-2-methyl-
P081 1,2,3-Propanetriol, trinitrate- (R)
P017 2-Propanone, 1-bromo-
P102 Propargyl alcohol
P003 2-propenal
P005 2-propen-1-ol
P067 1,2-Propylenimine
P102 2-Propyn-1-ol
P008 4-Pyridinamine
P075 Pyridine, (S)-3-(1-methyl-2-pyrrolidinyl)-, and salts
P111 Pyrophosphoric acid, tetraethyl ester

P103 Selenourea
P104 Silver cyanide
P105 Sodium azide
P106 Sodium cyanide
P107 Strontium sulfide
P108 Strychnidin-10-one, and salts
P018 Strychnidin-10-one, 2,3-dimethoxy-
P108 Strychnine and salts
P115 Sulfuric acid, thallium(I) salts

P109 Tetraethyldithiopyrophosphate
P110 Tetraethyl lead
P111 Tetraethylpyrophosphate
P112 Tetranitromethane (R)
P062 Tetraphosphoric acid, hexaethyl ester
P113 Thallic oxide
P113 Thallium(III) oxide
P114 Thallium (I) selenite
P115 Thallium (I) sulfate
P045 Thiofanax
P049 Thiomidodicarbonic diamide
P014 Thiophenol
P116 Thiosemicarbazide
P026 Thiourea, (2-chlorophenyl)-
P072 Thiourea, 1-naphthalenyl-
P093 Thiourea, phenyl
P123 Toxaphene
P118 Trichloromethanethiol

P119 Vanadic acid, ammonium salt
P120 Vanadium pentoxide
P120 Vanadium(V) oxide

P001 Warfarin, when present at concentrations greater than 0.3%

P121 Zinc cyanide
P122 Zinc phosphide, when present at concentrations greater than 10%

The following U code wastes are nonacutely hazardous.

U001 Acetaldehyde (I)
U034 Acetaldehyde, trichloro-
U187 Acetamide, N-(4-ethoxyphenyl)-
U005 Acetamide, N-9H-fluoren-2-yl-
U112 Acetic acid, ethyl ester (I)
U144 Acetic acid, lead salt
U214 Acetic acid, thallium(I) salt

U002   Acetone (I)
U003   Acetonitrile (I,T)
U248   3-(alpha-Acetonylbenzyl)-4-hydroxycoumarin and salts, when present at
       concentations of 0.3% or less
U004   Acetophenone
U005   2-Acetylaminofluorene
U006   Acetyl chloride (C,R,T,)
U007   Acrylamide
U008   Acrylic acid (I)
U009   Acrylonitrile
U150   Alanine, 3-[p-bis(2-chloroethyl)amino] phenyl-, L-
U328   2-Amino-I-methylbenzene
U353   4-Amino-I-methylbenzene
U011   Amitrole
U012   Aniline (I,T)
U014   Auramine
U015   Azaserine
U010   Azirino (2',3',3',4)pyrrolo (1,2-a)indole-4,7-dione, 6-amino-8-[((amino-
       carbonyl)oxy)methyl]-1,1a,2,8,8a,8b-hexahydro-8a-methoxy-5-
       methyl-

U157   Benz(j)aceanthrylene, 1,2-dihydro-3-methyl-
U016   Benz(c)acridine
U016   3,4-Benzacridine
U017   Benzal chloride
U018   Benz(a)anthracene
U018   1,2-Benzanthracene
U094   1,2-Benzanthracene, 7,12-dimethyl-
U012   Benzenamine (I,T)
U014   Benzenamine, 4,4'-carbonimidoylbis(N,N-dimethyl)-
U049   Benzenamine, 4-chloro-2-methyl-
U093   Benzenamine, N,N'-dimethyl-4-phenylazo-
U158   Benzenamine, 4,4'-methylenebis(2-chloro)-
U222   Benzenamine, 2-methyl-,hydrochloride
U181   Benzenamine, 2-methyl-,5-nitro
U019   Benzene (I,T)
U038   Benzeneacetic acid, 4-chloro-alpha-(4-chloro-phenyl)-alpha-hydroxy,
       ethyl ester
U030   Benzene, 1-bromo-4-phenoxy-
U037   Benzene, chloro
U190   1,2-Benzenedicarboxylic acid anhydride
U028   1,2-Benzenedicarboxylic acid [bis(2-ethyl-hexyl)] ester
U069   1,2-Benzenedicarboxylic acid, dibutyl ester
U088   1,2-Benzenedicarboxylic acid, diethyl ester
U102   1,2-Benzenedicarboxylic acid, dimethyl ester
U107   1,2-Benzenedicarboxylic acid, di-n-octyl ester
U070   Benzene, 1,2-dichloro-

U071	Benzene, 1,3-dichloro-
U072	Benzene, 1,4-dichloro-
U017	Benzene, (dichloromethyl)-
U223	Benzene, 1,3-diisocyanatomethyl- (R,T)
U239	Benzene, dimethyl- (I,T)
U201	1,3-Benzenediol
U127	Benzene, hexachloro-
U056	Benzene, hexahydro- (I)
U188	Benzene, hydroxy-
U220	Benzene, methyl-
U105	Benzene, 1-methyl-1,2,4-dinitro-
U106	Benzene, 1-methyl-2,6-dinitro-
U203	Benzene, 1,2-methylenedioxy-4-allyl-
U141	Benzene, 1,2-methylenedioxy-4-propenyl-
U090	Benzene, 1,2-methylenedioxy-4-propyl-
U055	Benzene, (1-methylethyl) (I)
U169	Benzene, nitro- (I,T)
U183	Benzene, pentachloro-
U185	Benzene, pentachloro-nitro-
U020	Benzenesulfonic acid chloride (C,R)
U020	Benzenesulfonyl chloride (C,R)
U207	Benzene, 1,2,4,5-tetrachloro-
U023	Benzene, (trichloromethyl)- (C,R,T)
U234	Benzene, 1,3,5-trinitro (R,T)
U021	Benzidine
U202	1,2-Benzisothiazolin-3-one,1,1-dioxide
U120	Benzo(j,k)fluorene
U022	Benzo(a)pyrene
U022	3,4-Benzopyrene
U197	p-Benzoquinone
U023	Benzotrichloride (C,R,T)
U050	1,2-Benzphenanthrene
U085	2,2'-Bioxirane (I,T)
U021	(1,1'-Biphenyl)-4,4'-diamine
U073	(1,1'-Biphenyl)-4,4'-diamine, 3,3'-dichloro-
U091	(1,1'-Biphenyl)-4,4'-diamine, 3,3'-dimethoxy-
U095	(1,1'-Biphenyl)-4,4'-diamine, 3,3'-dimethyl- U024 Bis(2-chloroethoxy) methane
U027	Bis(2-chloroisopropyl) ether
U244	Bis(dimethylthiocarbamoyl) disulfide
U028	Bis(2-ethyhexyl)phthalate (DEHP)
U246	Bromine cyanide
U225	Bromoform
U030	4-Bromophenyl phenyl ether
U128	1,3-Butadiene, 1,1,2,3,4,4-hexachloro
U172	1-Butanamine, N-butyl-N-nitroso-
U035	Butanoic acid, 4-[Bis(2-chloroethyl)amino]benzene-

U031 1-Butanol (I)
U159 2-Butanone (I,T)
U160 2-Butanone peroxide (R,T)
U053 2-Butenal
U074 2-Butene, 1,4-dichloro- (I,T)
U031 n-Butyl alcohol (I)

U136 Cacodylic acid
U032 Calcium chromate
U238 Carbamic acid, ethyl ester
U178 Carbamic acid, methylnitroso-, ethyl ester
U176 Carbamide, N-ethyl-N-nitroso-
U177 Carbamide, N-methyl-N-nitroso-
U219 Carbamide, thio-
U097 Carbamoyl chloride, dimethyl-
U215 Carbonic acid, dithallium (I)salt
U156 Carbonochloridic acid, methyl ester (I,T)
U033 Carbon oxyfluoride (R,T)
U211 Carbon tetrachloride
U033 Carbonyl fluoride (R,T)
U034 Chloral
U035 Chlorambucil
U036 Chlordane, technical
U026 Chlornaphazine
U037 Chlorobenzene
U039 4-Chloro-m-cresol
U041 1-Chloro-2,3-epoxypropane
U042 2-Chloroethyl vinyl ether
U044 Chloroform
U046 Chloromethyl methyl ether
U047 beta-Chloronaphthalene
U048 o-Chlorophenol
U049 4-Chloro-o-toluidine, hydrochloride
U032 Chromic acid, calcium salt
U050 Chrysene
U051 Creosote
U052 Cresols
U052 Cresylic acid
U053 Crotonaldehyde
U055 Cumene (I)
U246 Cyanogen bromide
U197 1,4-Cyclohexadienedione
U056 Cyclohexane (I)
U057 Cyclohexanone (I)
U130 1,3-Cyclopentadiene, 1,2,3,4,5,5-hexa- chloro- U058 Cyclophosphamide

U240 2,4-D, salts and esters

U059 Daunomycin
U060 DDD
U061 DDT
U142 Decachloro octahydro-1,3,4-metheno-2H-cyclobuta(c,d) pentalen-2-one
U062 Diallate
U133 Diamine (R,T)
U221 Diaminotoluene
U063 Dibenz(a,h)anthracene
U063 1,2:5,6-Dibenzanthracene
U064 1,2:7,8-Dibenzopyrene
U064 Dibenz(a,i)pyrene
U066 1,2-Dibromo-3-chloropropane
U069 Dibutyl phthalate
U062 S-(2,3-Dichloroallyl)diisopropylthiocarbamate
U070 o-Dichlorobenzene
U071 m-Dichlorobenzene
U072 p-Dichlorobenzene
U073 3,3'-Dichlorobenzidine
U074 1,4-Dichloro-2-butene (I,T)
U075 Dichlorodifluoromethane
U192 3,5-Dichloro-N-(1,1-dimethyl-2-propynyl)benzamide
U060 Dichloro diphenyl dichloroethane
U061 Dichloro diphenyl trichloroethane
U078 1,1-Dichloroethylene
U079 1,2-Dichloroethylene
U025 Dichloroethyl ether
U081 2,4-Dichlorophenol
U082 2,6-Dichlorophenol
U240 2,4-Dichlorophenoxyacetic acid, salts and esters
U083 1,2-Dichloropropane
U084 1,3-Dichloropropene
U085 1,2:3,4-Diepoxybutane (I,T)
U108 1,4-Diethylene dioxide
U086 N,N-Diethylhydrazine
U087 O,O-Diethyl-S-methyl-dithiophosphate
U088 Diethyl phthalate
U089 Diethylstilbestrol
U148 1,2-Dihydro-3,6-pyradizinedione
U090 Dihydrosafrole
U091 3,3'-Dimethoxybenzidine
U092 Dimethylamine (I)
U093 Dimethylaminoazobenzene
U094 7,12-Dimethylbenz(a)anthracene
U095 3,3'-Dimethylbenzidine
U096 alpha,alpha-Dimethylbenzylhydroperoxide (R)
U097 Dimethylcarbamoyl chloride
U098 1,1-Dimethylhydrazine

U099   1,2-Dimethylhydrazine
U101   2,4-Dimethylphenol
U102   Dimethyl phthalate
U103   Dimethyl sulfate
U105   2,4-Dinitrotoluene
U106   2,6-Dinitrotoluene
U107   Di-n-octyl phthalate
U108   1,4-Dioxane
U109   1,2-Dipheylhydrazine
U110   Dipropylamine (I)
U111   Di-N-propylnitrosamine

U001   Ethanal (I)
U174   Ethanamine, N-ethyl-N-nitroso-
U067   Ethane, 1,2-dibromo-
U076   Ethane, 1,1-dichloro-
U077   Ethane, 1,2-dichloro-
U114   1,2-Ethanediylbiscarbamodithioic acid
U131   Ethane, 1,1,1,2,2,2-hexachloro-
U024   Ethane, 1,1'-[methylenebis(oxy)]bis(2-chloro)-
U003   Ethanenitrile (I,T)
U117   Ethane, 1,1'-oxybis- (I)
U025   Ethane, 1,1'-oxybis(2-chloro)-
U184   Ethane pentachloro-
U208   Ethane, 1,1,1,2-tetrachloro-
U209   Ethane, 1,1,2,2-tetrachloro-
U218   Ethanethioamide
U247   Ethane, 1,1,1-trichloro-2,2-bis(p-methoxyphenyl)
U227   Ethane, 1,2,1-trichloro-
U043   Ethene, chloro-
U042   Ethene, 2-chloroethoxy-
U078   Ethene, 1,1-dichloro-
U079   Ethene, trans-1,2-dicloro-
U210   Ethene, 1,1,2,2-tetrachloro-
U173   Ethanol, 2,2'-(nitrosoimino)bis-
U004   Ethanone, 1-phenyl-
U006   Ethanoyl chloride (C,R,T)
U112   Ethyl acetate (I)
U113   Ethyl acrylate (I)
U238   Ethyl carbamate (urethan)
U038   Ethyl 4,4'-dichlorobenzilate
U359   Ethylene glycol monoethyl ether
U114   Ethylenebis(dithiocarbamic acid)
U067   Ethylene dibromide
U077   Ethylene dichloride
U115   Ethylene oxide (I,T)
U116   Ethylene thiourea

U117   Ethyl ether
U076   Ethylidene dichloride
U118   Ethylmethacrylate
U119   Ethyl methanesulfonate

U139   Ferric dextran
U120   Fluoranthene
U122   Formaldehyde
U123   Formic acid (C,T)
U124   Furan (I)
U125   2-Furancarboxaldehyde (I)
U147   2,5-Furandione
U213   Furan, tetrahydro- (I)
U125   Furfural (I)
U124   Furfuran (I)

U206   D-Glucopyranose,2-deoxy-2(3-methyl-3-nitro-soureido)-
U126   Glycidylaldehyde
U163   Guanidine, N-nitroso-N-methyl-N'nitro-

U127   Hexachlorobenzene
U128   Hexachlorobutadiene
U129   Hexachlorocyclohexane(gamma isomer)
U130   Hexachlorocyclopentadiene
U131   Hexachloroethane
U132   Hexachlorophene
U243   Hexachloropropene
U133   Hydrazine (R,T)
U086   Hydrazine, 1,2-diethyl-
U098   Hydrazine, 1,1-dimethyl-
U099   Hydrazine, 1,2-dimethyl-
U109   Hydrazine, 1,2-diphenyl-
U134   Hydrofluoric acid (C,T)
U134   Hydrogen fluoride (C,T)
U135   Hydrogen sulfide
U096   Hydroperoxide, 1-methyl-1-phenylethyl- (R)
U136   Hydroxydimethylarsine oxide

U116   2-Imidazolidinethione
U137   Indeno(1,2,3-cd)pyrene
U139   Iron dextran
U140   Isobutyl alcohol (I,T)
U141   Isosafrole

U142   Kepone

U143　Lasiocarpine
U144　Lead acetate
U145　Lead phosphate
U146　Lead subacetate
U129　Lindane

U147　Maleic anhydride
U148　Maleic hydrazide
U149　Malononitrile
U150　Melphalan
U151　Mercury
U152　Methacrylonitrile (I,T)
U092　Methanamine, N-methyl- (I)
U029　Methane, bromo-
U045　Methane, chloro- I,T)
U046　Methane, chloromethoxy-
U068　Methane, dibromo-
U080　Methane, dichloro-
U075　Methane, dichlorodifluoro-
U138　Methane, iodo-
U119　Methanesulfonic acid, ethyl ester
U211　Methane, tetrachloro-
U121　Methane, trichlorofluoro-
U153　Methanethiol (I,T)
U225　Methane, tribromo-
U044　Methane, trichloro-
U121　Methane, trichlorofluoro-
U123　Methanoic acid (C,T)
U036　4,7-Methanoindan, 1,2,4,5,6,7,8,8-octachloro-3a,4,7,7a- tetrahydro-
U154　Methanol (I)
U155　Methapyrilene
U247　Methoxychlor
U154　Methyl alcohol (I)
U029　Methyl bromide
U186　1-Methylbutadiene (I)
U045　Methyl chloride (I,T)
U156　Methyl chlorocarbonate (I,T)
U226　Methyl chloroform
U157　3-Methylcholanthrene
U158　4,4'-Methylenebis(2-chloroaniline)
U132　2,2'-Methylenebis(3,4,6-trichlorophenol)
U068　Methylene bromide
U080　Methylene chloride
U122　Methylene oxide
U159　Methyl ethyl ketone (I,T)
U160　Methyl ethyl ketone peroxide (R,T)
U138　Methyl iodide

U161   Methyl isobutyl ketone (I)
U162   Methyl methacrylate (I,T)
U163   N-Methyl-N'-nitro-N-nitrosoguanidine
U161   4-Methyl-2-pentanone (I)
U164   Methylthiouracil
U010   Mitomycin C

U059   5,12-Naphthacenedione,(8S-cis)-8-acetyl-10-[(3-amino-2,3,6-trideoxy-
       alpha-L-lyxo-hexopyranosyl)oxyl)-7,8,9,10-tetrahydro-6,8,11-trihy
       droxy-1-methyoxy-
U165   Naphthalene
U047   Naphthalene,2-chloro-
U166   1,4-Naphthalenedione
U236   2,7-Naphthalenedisulfonic acid,3,3'-[(3,3'-dimethyl-(1,1'bi-phenyl)-
       4,4'diyl)]-bis(azo)bis(5-amino-4-hydroxy)-, tetrasodium salt
U166   1,4,Naphthaquinone
U167   1-Naphthylamine
U168   2-Naphthylamine
U167   alpha-Naphthylamine
U168   beta-Naphthylamine
U026   2-Naphthylamine, N,N'-bis(2-chloromethyl)-
U169   Nitrobenzene (I,T)
U170   p-Nitrophenol
U171   2-Nitropropane (I)
U172   N-Nitrosodi-n-butylamine
U173   N-Nitrosodiethanolamine
U174   N-Nitrosodiethylamine
U111   N-Nitroso-N-propylamine
U176   N-Nitroso-N-ethylurea
U177   N-Nitroso-N-methylurea
U178   N-Nitroso-N-methylurethane
U179   N-Nitrosopiperidine
U180   N-Nitrosopyrrolidine
U181   5-Nitro-o-toluidine
U193   1,2-Oxathiolane,2,2-dioxide
U058   2H-1,3,2-Oxazaphosphorine,2-[bis(2-chloroethyl)amino] tetrahydro-,
       oxide 2-

U115   Oxirane (I,T)
U041   Oxirane, 2-(chloromethyl)-

U182   Paraldehyde
U183   Pentachlorobenzene
U184   Pentachloroethane
U185   Pentachloronitrobenzene
U186   1,3-Pentadiene (I)
U187   Phenacetin

U188	Phenol
U048	Phenol, 2-chloro-
U039	Phenol, 4-chloro-3-methyl-
U081	Phenol, 2,4-dichloro-
U082	Phenol, 2,6-dichloro-
U101	Phenol, 2,4-dimethyl-
U170	Phenol, 4-nitro-
U137	1,10-(1,2-phenylene)pyrene
U145	Phosphoric acid, Lead salt
U087	Phosphorodithioic acid O,O-diethyl-,S-methylester
U189	Phosphorous sulfide (R)
U190	Phthalic anhydride
U191	2-Picoline
U192	Pronamide
U194	1-Propanamine (I,T)
U110	1-Propanamine, N-propyl- (I)
U066	Propane, 1,2-dibromo-3-chloro-
U149	Propanedinitrile
U171	Propane, 2-nitro- (I)
U027	Propane, 2,2'-oxybis(2-chloro)-
U193	1,3-Propane sultone
U235	1-Propanol, 2,3-dibromo-,phosphate(3:1)
U126	1-Propanol, 2,3-epoxy-
U140	1-Propanol, 2-methyl- (I,T)
U002	2-Propanone (I)
U007	2-Propenamide
U084	Propene, 1,3-dichloro-
U243	1-Propene, 1,1,2,3,3,3-hexachloro-
U009	2-Propenenitrile
U152	2-Propenenitrile, 2-methyl- (I,T)
U008	2-Propenoic acid (I)
U113	2-Propenoic acid, ethyl ester (I)
U118	2-Propenoic acid, 2-methyl-, ethyl ester
U162	2-Propenoic acid, 2-methyl, methyl ester (I,T)
U194	n-Propylamine (I,T)
U083	Propylene dichloride
U196	Pyridine
U155	Pyridine, 2-[(2-(dimethylamino)-2-thenylamino)]
U179	Pyridine, hexahydro-N-nitroso-
U191	Pyridine, 2-methyl-
U164	4(1H)-Pyrimidinone, 2,3-dihydro-6-methyl-2-thioxo-
U180	Pyrrole, tetrahydro-N-nitroso-
U200	Reserpine
U201	Resorcinol
U202	Saccharin and salts

U203 Safrole
U204 Selenious acid
U204 Selenium dioxide
U205 Selenium disulfide (R,T)
U015 L-Serine, diazoacetate (ester)
U089 4,4'-Stilbenediol,alpha,alpha'-diethyl-
U206 Streptozotocin
U135 Sulfur hydride
U103 Sulfuric acid, dimethyl ester
U189 Sulfur phosphide (R)
U205 Sulfur selenide (R,T)

U207 1,2,4,5-Tetrachlorobenzene
U208 1,1,1,2-Tetrachloroethane
U209 1,1,2,2-Tetrachloroethane
U210 Tetrachloroethylene
U213 Tetrahydrofuran (I)
U214 Thallium(I) acetate
U215 Thallium(I) carbonate
U216 Thallium(I) chloride
U217 Thallium(I) nitrate
U218 Thioacetamide
U153 Thiomethanol (I,T)
U219 Thiourea
U244 Thiram
U220 Toluene
U221 Toluenediamine
U223 Toluenediisocyanate (R,T)
U328 o-Toluidine
U222 o-Toluidine hydrochloride
U353 p-Toluidine
U011 1H-1,2,4-Triazol-3-amine
U226 1,1,1-Trichloroethane
U227 1,1,2-Trichloroethane
U228 Trichloroethene
U228 Trichloroethylene
U121 Trichloromonofluoromethane
U234 sym-Trinitrobenzene (R,T)
U182 1,3,5-Trioxane, 2,4,5-trimethyl-
U235 Tris(2,3-dibromopropyl)phosphate
U236 Trypan blue

U237 Uracil, 5[bis(2-chloromethyl)amino]-
U237 Uracil mustard

U043 Vinyl chloride

U248   Warfarin, when present at concentrationsof 0.3% or less

U239   Xylene (I)

U200   Yohimban-16-carboxylic acid, 11,17-dimethoxy-18-[(3,4,5 trimethoxy-benzoyl)oxy]-methyl ester

U249   Zinc phosphide, when present at concentrations of 10% or less

APPENDIX F

Incompatible Chemicals in Storage and in Reactions

Acetic Acid - with chromic acid, ethylene glycol, hydroxyl-containing compounds, nitric acid, perchloric acid, permanganates, peroxides, etc.

Acetone - with concentrated nitric and sulfuric acid mixtures, etc.

Acetylene - with copper tubing, halogens (bromine, chlorine, fluorine, and iodine), mercury, silver and their compounds, etc.

Alkali Metals - with carbon dioxide, carbon tetrachloride, chlorinated hydrocarbons, water, etc.

Ammonia, Anhydrous - with calcium hypochlorite, halogens (bromine, chlorine, fluorine, and iodine), hydrogen fluoride, mercury, etc.

Ammonium Nitrate - with acids, chlorates, flammable liquids, metal powders, nitrates, sulfur, finely divided organic compounds and combustibles, etc.

Aniline - with hydrogen peroxide, nitric acid, etc.

Bromine - with acetylene, ammonia, butadiene, butane, hydrogen, sodium carbide, turpentine, metal powders, etc.

Carbon (Activated Charcoal) - with calcium hypochlorite, oxidizing agents, etc.

Chlorates - with acids, ammonium salts, carbon, metal powders, sulfur, finely divided organic compounds and combustibles, etc.

Chlorine - with acetylene, ammonia, benzine and other petroleum fractions, butadiene, hydrogen, sodium carbide, turpentine, metal powders, etc.

Chlorine Dioxide - with ammonia, hydrogen sulfide, methane, phosphine, etc.

Chromic Acid - with acetic acid, alcohol, camphor, glycerine, naphthalene, turpentine, flammable liquids, etc.

Copper - with acetylene, hydrogen peroxide, etc.

Cyanides - with acids, alkalis, etc.

Flammable Liquids - with ammonium nitrate, chromic acid, halogens (bromine, chlorine, fluorine, and iodine), hydrogen peroxide, nitric acid, sodium peroxide, etc.

Hydrocarbons - with chromic acid, halogens (bromine, chlorine, fluorine, and iodine), sodium peroxide, etc.

Hydrogen Peroxide - with aniline, copper, chromium, iron, most metals, nitromethane, flammable liquids, combustible materials, etc.

Hydrogen Sulfide - with fuming nitric acid, oxidizing gases, etc.

Iodine - with ammonia, acetylene, etc.

Mercury - with acetylene, hydrogen, fulminic acid, etc.

Nitric Acid - with acetic acid, aniline, carbon (activated), chromic acid, hydrocyanic acid, hydrogen sulfide, organic compounds that are readily nitrated, etc.

Oxalic Acid - with mercury, silver, etc.

Oxygen - with grease, hydrogen, oils, flammable materials, etc.

Perchloric Acid - with acetic anhydride, alcohol, bismuth and its alloys, flammable and combustible materials, etc.

Phosphorus Pentoxide - with water.

Potassium Permanganate - with benzaldehyde, ethylene glycol, glycerine, sulfuric acid, etc.

Silver - with acetylene, ammonium compounds, oxalic acid, tartaric acid, etc.

Sodium - with carbon dioxide, carbon tetrachloride, water, etc.

Sodium Peroxide - with acetic anhydride, glacial acetic acid, benzaldehyde, carbon disulfide, ethyl acetate, ethylene glycol, furfural, glycerine, methanol, any oxidizable chemicals, reducing agents, etc.

Sulfuric Acid - with chlorates, perchlorates, permanganates, water, etc.

INDEX

A

ABCs, *see* Airway. breathing, and circulation
Absorbents, 145–146, *see also* specific types
Absorbers, 112, *see also* specific types
Absorption of chemicals, 75, 141
Absorption spectrophotometry, 115
Access to records, 196–197
Accidents, 21, 26, 28, 125, 133, *see also* specific types
    response to, *see* Emergency response
Acetal, 69
Acetaldehyde, 37, 53, 62
Acetic acid, 37, 62, 83, 154
Acetic anhydride, 154
Acetone, 37, 60, 62, 65
Acetonitrile, 62
2-Acetylaminofluorene, 10
Acetylene, 64, 68, 71
Acetylenic compounds, 68, *see also* specific types
Acetylides, 67, *see also* specific types
ACGIH, *see* American Conference of Governmental Industrial Hygienists
Acids, 82, 133, *see also* specific types
    emergency response procedures for, 154
    leaks of, 154
    storage of, 84, 87, 88
    strong, 72
Acinolite, 11
Acrolein, 67
Acrylic acid, 69
Acrylonitrile, 10, 37, 69
Active exposure monitoring, 110–112
Acutely hazardous wastes, 180, 181

Add-air hoods, 101
Administrative unit supervisor, 15
Adsorption, 145–146
Aerosols, 53, 61, *see also* specific types
AFFF, *see* Aqueous film forming foam
Air cleaners, 93
Air contaminants, 106, 110, 113, 137–140, *see also* specific types
Air-line (supplied-air) respirators, 42
Air monitoring, 107–110, 144
Air-moving devices, 110
Air-purifying respirators (APRs), 40–42
Air sampling, 140
Airway, breathing, and circulation (ABCs), 167, 169–170
Alarm system, 128–129
Alcohols, 51, 62, 65, 84, *see also* specific types
Aldehydes, 10, 11, 37, 53, 62, 84, *see also* specific types
Aliphatic hydrocarbons, 70, *see also* specific types
Alkali metals, 151
Allergic response, 77–78
Aluminum alkyls, 71, 87
Aluminum bromide, 71
American Conference of Governmental Industrial Hygienists (ACGIH), 5, 14, 109, 141, 174
American National Standards Institute (ANSI), 5, 23, 43, 140
American Society of Mechanical Engineers (ASME), 5
Amides, 84, *see also* specific types
Amines, 84, *see also* specific types

251

4-Aminodiphenyl, 10
Ammonia, 64, 77, 154
Ammonium hydroxide, 37, 53
Ammonium nitrate, 49, 50, 151
Anemometers, 98
Anhydrides, 84, 154, *see also* specific types
Aniline, 37, 62
ANSI, *see* American National Standards
    Institute
Anthophyllite, 11
Anthrax, 52
Aprons, 35, 73, 74
APRs, *see* Air-purifying respirators
Aqueous film forming foam (AFFF), 150
Area monitoring, 108
Arsenates, 84, *see also* specific types
Arsenic, 10, 183
Asbestos, 11
ASME, *see* American Society of Mechanical
    Engineers
Asphyxiation, 76
Atomic absorption spectrophotometry, 115
Atomic Energy Act of 1954, 1
Atropine, 175
Autoignition temperature, 60, 61
Azides, 67, 68, 84, *see also* specific types
Azo compounds, 68, *see also* specific types
Azomethane, 68

B

Barium, 183
Barium oxide, 53
Barium sulfate, 77
Barometric pressure, 106–107
Baseline profile of laboratory workers, 120
Bases, 72, 82, 88, 154, *see also* specific
    types
Benzaldehyde, 37
Benzene, 11, 37, 62, 76, 78, 183
Benzenesulfonyl azide, 68
Benzidine, 11
Benzoyl chloride, 53
Benzvalene, 68
Benzyl chloride, 37
Beryllium, 78
Bis(chloromethyl)ether, 11
Black powder, 49
Blasting agents, 49–50, 61, *see also* specific
    types
Blasting caps, 49
Body protection, 34–35
Boiling point, 61

Bone oil, 53
Boots, 156, 159–161
Borates, 84, *see also* specific types
Boron tribromide, 71
Botulism, 52
Breathing equipment, *see* Respirators;
    Respiratory protection
Bromine, 37, 73
Butadiene, 68, 69
Butane, 37
Butylaldehyde, 37
Bypass-air hoods, 100–101

C

Cadmium, 183
Calcium, 51, 65
Calcium carbide, 71
Calcium carbonate, 77
Calcium hypochlorite, 37
Calcium oxide, 71, 72, 157
Calibration, 116
Cancer, 78, *see also* Carcinogens
Capture distance, 98
Capturing hoods, 96–97
Carbides, 71, 84, *see also* specific types
Carbolic acid, 49
Carbonates, 77, 84, 157, *see also* specific
    types
Carbon dioxide, 51, 77
Carbon dioxide fire extinguishers, 45
Carbon disulfide, 37, 62
Carbon monoxide, 64, 74, 76
Carbon monoxide absorbers, 42
Carbon tetrachloride, 37, 49, 151, 183
Carcinogens, 9, 23, 78, 89, *see also* Cancer;
    specific types
Cardiopulmonary resuscitation (CPR), 122,
    168–169
Cellular damage, 77
Cements, 147
Cervical immobilization, 167
Cesium trioxide, 71
Chain-of-command, 133–134, 143
Chain of hazard evaluation, 127
Chemical compatibility, 82–84
Chemical Emergency Officer, 125–126, 132
Chemical emergency response, *see*
    Emergency response
Chemical handling procedures, *see* Handling
    procedures
Chemical Hygiene Officer, 15–16, 22, 44, 130
Chemical hygiene plan, 15–16, 19–23

Chemical inactivation, 157–158
Chemically impregnated test media, 10
Chemical storage, 81–89
  for carcinogens, 89
  compatibility and, 82–84
  for corrosives, 88
  for flammable chemicals, 86–88
  for highly toxic chemicals, 89
  inventory control and, 81–82, 89
  for oxidizers, 87, 88
  storerooms for, 84–85
  for toxic chemicals, 88–89
  ventilation for, 88
  for water-reactive chemicals, 87–88
Chief executive officer, 15
Chlorates, 52, 84, 151, see also specific
    types
Chlordane, 183
Chlorine, 37, 50, 51, 55
  emergency response procedures for, 153,
    154
  handling procedures for, 67, 73, 77
Chlorine Institute, 153
Chlorites, 84, see also specific types
Chloroacetone, 37
Chloro compounds, 68, see also specific
    types
Chloroform, 37, 183
Chloroprene, 69
Chlorotrifluoroethylene, 69
Chromates, 84, see also specific types
Chromatography, 115, see also specific
    types
Chromic acid, 37, 73
Chromium, 53, 78, 183
Chronic pulmonary disease, 77
Chronic toxicity, 78
Cleanup, 28, 74, 182–183
Closed-circuit SCBA, 43, 44
Clothing, 34–35, 74, 156, see also specific
    types
  clean, 163
  in emergency response, 134, 135, 153
    clean, 163
    removal of, 162–163
  heat stress and, 172
  removal of, 162–163
Coal Mine Health and Safety Act of 1969, 1
Coal tar pitch volatiles, 11
Coke oven emissions, 11
Colorimetric indicator tubes, 114–115, 138
Colorimetry, 139
Combustible gas meters, 139

Combustibles, 51, 55, 60, see also
    Flammable; specific types
Communication, 128, 142
Compatibility of chemicals, 82–84
Compressed gas, 42, 50–51
Compressed Gas Association, 43
Conditionally exempt small quantity
    generators, 185
Consciousness level assessment, 170
Contact lenses, 33
Containers, 184
Containment, 145–147
Contaminant dispersion pattern, 98
Contaminant escape pattern, 98
Contamination avoidance, 155–156
Contamination minimization, 155–156
Contamination reduction (limited access)
    zone, 136, 159
Continuous (integrated) samples, 108, 111
Copper acetylide, 68
Copper chloride, 53
Copper fulminate, 68
Corrosives, 9, 48, 50, 53, see also specific
    types
  classes of, 72–73
  defined, 153
  emergency response procedures for, 153–
    154
  handling procedures for, 71–74
  placards for, 55
  storage of, 88
Corrosivity, 180, 181
Cotton dust, 11
CPR, see Cardiopulmonary resuscitation
Cresols, 84, 183, see also specific types
Critical care life support, 165
Cryogenic hazards, 29
Cumene, 69
Cyanides, 84, see also specific types
Cyanogen, 51
Cyclohexane, 37, 62
Cyclohexene, 69

D

Data handling, 116
Decontamination, 131
  defined, 155
  effectiveness of, 158
  of emergency responders, 155–163
    effectiveness of, 158
    methods in, 157–158
    planning for, 156–157

methods in, 157–158
planning for, 156–157
of samples, 158–163
of victims, 170–172
Deflecting vane velometers, 98
Dehydrating agents, 72–73, 88, *see also*
    specific types
De minimis violation under OSHAct, 4
Department of Defense, 1
Department of Transportation (DOT), 18,
    43, 47
    emergency planning and, 132
    emergency response and, 152, 153
    Emergency Response Guide of, 48
    hazardous chemical transport guidelines
        of, 48–54
    placarding/labeling system of, 54, 55
    waste management and, 186
Derivative spectroscopy, 115
Design standards of OSHA, 6
Diacetylene, 69
Diacetyl peroxide, 68
Diacyl peroxides, 70
Diazomethane, 68
Dibenzyl ether, 37
Diborane, 65
1,2-Dibromo-3-chloropropane, 11
Dibutyl phthalate, 37
1,4-Dichlorobenzene, 183
3,3'-Dichlorobenzidine, 11
1,2-Dichloroethane, 183
1,1-Dichloroethylene, 183
Dichloromethylsilane, 65
Diethanolamine, 37
Diethyl ether, 37, 62, 69
Diethylethoxyaluminum, 65
Diffusion, 110, 113
Diffusion devices, 114
Dilution, 154
Dilution ventilation, 91–92
4-Dimethylaminoazobenzene, 11
Dimethylformamide, 62
2,4-Dinitrotoluene, 183
Dioxane, 69
Dip-and-read tests, 10
Direct-reading instruments, 138–139
Discarded commercial chemical products,
    181
Disopropyl ether, 69
Dispersion pattern, 98
Divinyl acetylene, 69
Divinyl ether, 65, 69
Documentation, 132, 145

Dosage, 74–75
Dosimeters, 112, 114
DOT, *see* Department of Transportation
Drifts, 148
Dry chemical fire extinguishers, 45
Dry chemicals, 150, *see also* specific types
Dry earth, 151
Dry ice, 53
Dry inert material, 151
Dry powdered limestone, 151
Dry powder fire extinguishers, 45
Dry salt lime, 151
Dry soda ash, 151
Ducts, 93, 99
Dusts, 11, 72, 77, 106, 134, 151, *see also*
    specific types
Dynamite, 49

E

Earth metals, 71, *see also* specific types
Electrical fires, 45, 46
Electronic circuitry, 138
Electron spectroscopy, 115
Emergencies, *see also* specific types
    documentation of, 132, 145
    multiple, 134
    potential, 126–127
    response to, *see* Emergency response
    training for, 131–132
    victim handling in, *see* Victim handling
Emergency evacuation, *see* Evacuations
Emergency life-saving procedures, 171
Emergency lighting, 129
Emergency medical technicians (EMTs),
    122
Emergency medical treatment, 121–122, 131
Emergency Officer, 125–126, 132
Emergency respirators, 39
Emergency responders, 133, 137, 154, *see*
    also Emergency response
    clothing for, 156, 157, 159, 160–163
    communication to, 128
    decontamination of, 155–163
        effectiveness of, 158
        methods in, 157–158
        planning for, 156–157
    first, 141–142
    medical examination of, 163
Emergency response, 26, 133–154, *see also*
    Emergencies; Emergency responders
    alarm system and, 128–129
    chain-of-command for, 133–134, 143

clothing in, 134, 135, 153, 156, 157, 160–163
  clean, 163
  removal of, 162–163
communication in, 128, 142
containment in, 145–147
for corrosives, 153–154
documentation of, 132, 145
equipment for, 26, 40, 44–46, 130, 133, 157, *see also* specific types
exposure levels and, 140–141
first responder in, 141–142
for flammable chemicals, 149–151
follow-up to, 132
hazard area control in, 135–137, 142–143
hazard exposure assessment in, 137–140
initial response action in, 142–145
for organic peroxides, 152
for oxidizers, 151–152
for peroxides, 151, 152
personnel protection in, 134–135
planning for, 17, 125–132
for poisons, 152–153
procedures for, 21, 127–128
review of, 132
site characterization for, 143–144
stages of, 165–166
supplies for, 133
for toxic chemicals, 152–153
triage in, 167–169
victim handling in, *see* Victim handling
Emergency Response Guide of DOT, 48
Emergency telephone numbers, 27
Employees, *see* Laboratory workers
Employee training, *see* Training
Employer's responsibilities under OSHAct, 13–18
Empty containers, 184
EMTs, *see* Emergency medical technicians
Enclosure hoods, 94–95
Endothermic compounds, 67, *see also* specific types
Endothermic reactions, 66
Endrin, 183
Energy of activation, 66, 67
Entry-and-escape SCBA, 43–44
Environmental Protection Agency (EPA), 18, 47, 134, 143
  emergency planning and, 132
  waste management and, 179–182, 188
EP, *see* Extraction procedure
EPA, *see* Environmental Protection Agency
Epidemiology, 105

Epoxides, 183, *see also* specific types
Epoxy, 84, 147
Equipment, *see also* specific types
  emergency response, 26, 40, 44–46, 130, 133, 157
  inspection of, 31
  maintenance of, 26, 31
  mechanical, 29–30
  personal protective, *see* Personal protective equipment
  safety, 44–46
  selection of, 31
  storage of, 31, 130, 133
Escape-only SCBA, 43
Escape pattern, 98
Esters, 84, *see also* specific types
Ethers, 65, 84, *see also* specific types
Ether solvents, 69, *see also* specific types
Ethoxyacetylene, 68
Ethyl acetate, 37, 62
Ethyl alcohol, 51, 62
Ethylene dichloride, 37
Ethylene glycol, 37
Ethylene glycol dimethyl ether, 69
Ethyleneimine, 11
Ethylene oxide, 11, 62
Ethylene trichloride, 37
Etiologic agents, 52, *see also* specific types
Evacuations, 122, 129–130, 143, 153, 172
Excelsior, 53
Exclusion (restricted) zone, 135–136, 159
Exothermic reactions, 66
Explosimeters, 139
Explosions, 21, 60, 68, 133
Explosives, 48–50, 61, 67, *see also* specific types
  defined, 62
  placards for, 54, 55
Exposure, *see also* specific types
  absorption, 75, 141
  acute, 75, 76
  assessment of, 137–140
  chronic, 75, 76
  duration of, 74–75
  ingestion, 75, 176–177
  inhalation, 75, 77, 175–176
  levels of, 140–141
  minimization of, 14
  monitoring of, *see* Exposure monitorin
  protection from, 21
  record keeping and, 195
  routes of, 75, *see also* specific types
  to toxic chemicals, 74–75

Exposure monitoring, 105–116
  active, 110–112
  considerations in, 107–110
  criteria for, 107
  devices for, 110–115
  environmental factors affecting, 106–107
  interpretation of results of, 115–116
  objectives of, 105, 109
  passive, 110, 112–114
  procedures in, 110–115
  program for, 16–17
Extraction procedure (EP) toxicity test, 182
Eye contamination first aid, 176
Eye protection, 32–34, 44–45, 73–74
Eyewash fountains, 44–45, 73–74

F

Facepieces, 42, 141, 162
Face protection, 32–34, 141
Face velocity of fume hoods, 102–104
Fans, 93
FDA, *see* Food and Drug Administration
Ferric fluoride, 53
Ferrous sulfate, 53, 70
FID, *see* Flame ionization detectors
Filters, 42, 102, *see also* specific types
Fire extinguishers, 44–46, 150, 151
Fire-fighting foams, 146, 150, 151
Fire-hazard chemicals, *see* Flammable
  chemicals
Fire point, 60
Fire prevention, 60, 65
Fireworks, 49
First aid, 175–177
First aid station, 163
First responder, 141–142, *see also*
  Emergency responders
Fishcrap, 53
Flame ionization detectors, 139
Flammability, 57, 60–61, 62, *see also*
  Flammable chemicals
Flammable aerosols, 61
Flammable chemicals, *see also* Flammabil-
  ity; Flammable liquids; Flammable
  solids;
    specific types
  emergency response for, 149–151
  gaseous, 50, 51, 55, 57, 61, 64, 149
  handling procedures for, 60–64
  liquid, *see* Flammable liquids
  OSHA definitions of, 61–63

  physical properties, 60–61
  solid, *see* Flammable solids
  storage of, 82, 86–87, 88
Flammable gases, 50, 51, 55, 57, 61, 64,
  149, *see also* Flammable chemicals;
  specific types
Flammable liquids, 48, 50, 51, 57, 61, 65,
  see also Flammable chemicals;
  specific types
  defined, 61
  emergency response procedures for,
  149–150
  flash points for, 60
  properties of, 62
  vapor of, 64
Flammable solids, 48, 51–52, 55, *see also*
  Flammable chemicals; specific types
  defined, 61–62
  emergency response procedures for,
  150–151
  placards for, 54
Flashback, 61
Flash point, 60–63, 149
Fluorescence spectrophotometry, 115
Fluorine, 37, 65, 73
Foams, 146, 150, 151
Fog line dispersion, 146
Fogs, 106, 150
Food and Drug Administration (FDA), 47
Formaldehyde, 10, 11, 37
Formic acid, 37
Fourier transform infrared (FTIR)
  spectroscopy, 115
Freons, 151
FTIR, *see* Fourier transform infrared
Fuel oil, 50, 51
Fuels, 49, *see also* specific types
Full-compliance generators, 184
Full-facepiece masks, 42, 141
Fulminates, 67, 68, *see also* specific types
Fume hoods, 99–104
Fumes, 106, 154, *see also* specific types
Fuses, 49

G

Gases, 48, 50, 112, 113, *see also* specific
  types
  compressed, 42, 50–51
  defined, 106
  detection of, 138, 139
  flammable, 50, 51, 55, 57, 61, 64, 149

irritant, 77
    mixtures of, 109
    monitoring of, 110
    nonflammable, 50, 51, 55
    poison, 50, 51, 54, 55, 152, 153
    protection against, 134
    sampling of, 109
Gasoline, 49, 60, 62, 65
General safety practices, 25–30, *see also*
    specific types
General safety rules, 26–27
General ventilation (dilution ventilation),
    91–92
Generators of hazardous wastes, 184–185, 188
Germane, 65
Germanium chloride, 71
Glacial acetic acid, 83, 154
Glasgow Coma Scale, 170
Glassware safety rules, 27–28
Gloves, 35–38, 73, 74
    for emergency responders, 156, 159, 160
    removal of, 161, 162
    washing of, 160, 162
Glycerol, 37
Glycols, 84, *see also* specific types
Goggles, 33–34
Grab (instantaneous) samples, 108, 111
Gravimetric analysis, 115

H

Hair spray, 53
Half masks, 42
Halides, 71, 84, *see also* specific types
Halogenated hydrocarbons, 84, *see also*
    specific types
Halogens, 73, 84, *see also* specific types
Halon fire extinguishers, 45
Handling procedures, 59–79, *see also*
    specific types
    for corrosives, 71–74
    for flammable chemicals, 60–64
    for hypergolic mixtures, 65–66
    for pyrophoric chemicals, 64–65
    for reactive chemicals, 66–71
    for toxic chemicals, 74–79
    for water-reactive chemicals, 66, 70–71
Hand protection, 35–38
Hard hats, 162
Hazard area control, 135–137, 142–143
Hazard Communication Standard, 10, 17,
    196

Hazardous chemicals, *see also* Hazardous
    wastes; Hazards; specific types
    defined, 9
    degree 0, 56
    degree 1, 56
    degree 2, 56
    degree 3, 56
    degree 4, 55–56
    DOT system for identification of, 48–53,
        54
    flammability of, 57, *see also* Flammable
        chemicals
    health, 55–56
    identification of, 47–58
        DOT system of, 48–53, 54
        NFPA system for, 54–58
        United Nations numbering systems for,
            47–48
    NFPA system for identification of, 54–58
    reactivity of, 57–58
    regulation of, 47
    United Nations numbering systems for,
        47–48
Hazardous and Solid Waste Amendments
    (HSWA) to RCRA, 179, 188
Hazardous wastes, *see also* specific types
    acutely, 180, 181
    characteristics of, 180–182, *see also*
        specific types
    defined, 180–184
    EPA-listed, 180–181
    generators of, 184–185, 188
    management of, 18, 179–188
        defined, 180–184
        generators and, 184–185, 188
        on-site storage and, 185–186
        reduction of waste and, 188
        shipment and, 186–188
        uniform hazardous waste manifest in,
            187–188
    mixed, 182
    on-site storage of, 185–186
    reduction in, 188
    shipment of, 186–188
    storage of, 185–186
    toxic, 180
Hazards, *see also* Hazardous chemicals;
    Hazardous wastes; specific types
    assessment of, 31
    chain of evaluation of, 127
    control measures for, 19–20
    cryogenic, 29

determination of, 17
evaluation of, 127
exposure assessment for, 137–140
health, 9, 55–56
identification of, 142
special, 58
zones of, 135–137
Health hazards, 9, 55–56
Heated wire anemometers, 98
Heat exhaustion, 173
Heat rash, 173
Heat stress, 172–175
Heat stroke, 173–174
Heat syncope, 173
Helium, 51
Hematopoietic toxins, 9, *see also* specific
    types
Hemorrhage, 167
HEPA, *see* High-efficiency particulate
Hepatotoxins, 9, *see also* specific types
Heptachlor, 183
Hexachlorobenzene, 183
Hexachloro-1,3-butadiene, 183
Hexane, 37, 62, 139
High-efficiency particulate (HEPA) filters,
    102
Highly toxic chemicals, 9, 23, 89, *see also*
    specific types
High-pressure liquid chromatography
    (HPLC), 115
Hoods
    defined, 93
    emergency responders, 156
    fume, 99–104
    performance of, 98
    testing of, 97–99
    training in use of, 103–104
    types of, 93–97
    velocity measurements for, 98
Horizontal-type standards of OSHA, 6
Housekeeping safety rules, 28
HPLC, *see* High-pressure liquid chromatog-
    raphy
HSWA, *see* Hazardous and Solid Waste
    Amendments
Humidity, 106
Hydrazine, 62
Hydrides, 84, *see also* specific types
Hydrobromic acid, 38
Hydrocarbons, 70, 84, *see also* specific
    types
Hydrochloric acid, 38, 49, 77, 154

Hydrocyanic acid, 51
Hydrofluoric acid, 38
Hydrogen, 51, 64, 65, 67, 70, 87
Hydrogen peroxide, 38, 52, 67, 151
Hydrogen sulfide, 64
Hydroperoxides, 68, 69, 84, *see also* specific
    types
Hydroxides, 72, 84, *see also* specific types
Hypergolic mixtures, 65–66, *see also*
    specific types
Hypersensitivity, 118
Hypochlorites, 84, *see also* specific types

I

IARC, *see* International Agency for
    Research on Cancer
ICP, *see* Inductively coupled plasma
IDLH, *see* Immediately dangerous to life
    and health
Ignitability, 180, 181
Imides, 84, *see also* specific types
Imines, 84, *see also* specific types
Immediately dangerous to life and health
    (IDLH), 40, 43, 140–141, 144
Imminent danger, 4
Inactivation of chemicals, 157–158
Inductively coupled plasma (ICP), 115
Inert dusts, 77
Infrared radiation, 138
Infrared spectroscopy, 115
Ingestion of chemicals, 75, 176–177
Inhalation of chemicals, 75, 77, 175–176
Initial response action, 142–145
Injuries, 127, 167, *see also* specific types
Inorganic metal salts, 78, *see also* specific
    types
Inorganic peroxides, 52, *see also* specific
    types
Inspection of equipment, 31
Instantaneous (grab) samples, 108, 111
Insulated gloves, 36–38
Integrated (continuous) samples, 108, 111
International Agency for Research on
    Cancer (IARC), 23
Inventory control, 81–82, 89
Iodine, 38, 53
Ion chromatography, 115
Irritants, 9, 52, 54, 77, 152, *see also* specific
    types
Irritation, 77
Isocyanates, 84, *see also* specific types

J

Jumpsuits, 35

K

Kerosene, 51
Ketenes, 84, *see also* specific types
Ketones, 38, 62, 65, 69, 84, 183, *see also* specific types

L

Labeling, 17, 54, 55, *see also* specific types
Laboratory analysis of samples, 138–140
Laboratory clothing, *see* Clothing
Laboratory coats, 35, 73
Laboratory equipment, *see* Equipment
Laboratory glassware safety rules, 27–28
Laboratory Standard of OSHA, 9–11, 14
  exemptions from, 10–11
  explosives defined under, 62
  flammable chemicals defined under, 62
  laboratory worker information and, 20–21
  reactive chemicals under, 66
  record keeping under, 18, 123, 194–198
  training and, 20–21, 189, 194
Laboratory supervisors, 15, 16, 25, 130, 191, 193
Laboratory ventilation, *see* Ventilation
Laboratory waste management, *see* Hazardous wastes, management of
Laboratory workers, 16
  baseline profile of, 120
  defined, 10
  emergency medical treatment for, 121–122, 131
  in emergency response, *see* Emergency responders
  exposure monitoring of, *see* Exposure monitoring
  fitness-to-work of, 120
  information for, 20–21, 197, 198
  medical consultation for, 21–22, 117–124
    periodic, 120–121
    program development for, 117–118
  medical examination of, 21–22, 117–124
    periodic, 120–121
    pre-employment, 118–120
    program development for, 117–118
    at termination, 124
  medical history of, 118
  medical records of, 123, 195–197
  medical testing of, 121
  medical treatment for, 117, 121–123, 131
  monitoring of, 108
  nonemergency medical treatment for, 122–123
  occupational history of, 118
  physical examination of, 118–120
  pre-employment medical examination of, 118–120
  training of, *see* Training
  transferred, 194
  X-rays of, 196
"Lab pack", 186, 187
Lauroyl peroxide, 49
LC, *see* Lethal concentration
LD, *see* Lethal dose
Lead, 11, 76, 183
Lead azide, 68
Lead perchlorate, 68
Lead picrate, 68
Lead wool, 148
Leaks, 21, 64, 153, 154
Leak-stopping devices, 147–149
Leather gloves, 36
LELs, *see* Lower explosive limits
Lethal concentration (LC), 75
Lethal dose (LD), 74, 75
LFLs, *see* Lower flammability limits
Life-saving procedures, 171
Life support, 165
Life-threatening injuries, 167
Lime, 151, 157
Limestone, 151
Limited access (contamination reduction) zone, 136, 159
Lindane, 183
Liquid alcohols, 65, *see also* specific types
Liquid fluorine, 65
Liquid peroxides, 152, *see also* specific types
Liquids, 145–147, *see also* specific types
  flammable, *see* Flammable liquids
Lithium, 45, 52, 151
Lithium aluminum hydride, 45, 65
Local exhaust, 91–93
Lower explosive limits (LELs), 139
Lower flammability limits (LFLs), 60, 61, 149

M

Magnesium, 45, 65, 71
Magnitude of injury or damage, 127

Maintenance of equipment, 26, 31
Malathion, 52
Manganates, 84, *see also* specific types
Maritime Standard of OSHA, 139
Masks, 42, 141
Matches, 51
Material safety data sheets (MSDSs), 17, 21,
    73, 196
Mechanical equipment, 29–30, *see also*
    specific types
Medical consultation, 21–22, 117–124
    periodic, 120–121
    program development for, 117–118
Medical evaluation, 31
Medical examination, 21–22, 117–124
    of emergency responders, 163
    periodic, 120–121
    pre-employment, 118–120
    program development for, 117–118
    at termination, 124
Medical history, 118
Medical program review, 123
Medical records, 123, 195–197
Medical tests, 121
Medical treatment, 117, 121–123, 131, *see*
    also specific types
Mercury, 53, 183
Mercury fulminate, 68
Metal alkyls, 46, *see also* specific types
Metal fires, 45, 46
Metal halides, 71, *see also* specific types
Metal hydrides, 46, *see also* specific types
Metal hydroxides, 72, *see also* specific types
Metallic calcium, 51
Metallic mercury, 53
Metal oxides, 71, *see also* specific types
Metals, 45, 46, 67, 71, 78, 84, 151, *see also*
    specific types
Metal salts, 78, *see also* specific types
Methane, 51, 139
Methoxychlor, 183
Methyl acetylene, 69
Methyl alcohol, 62
Methylamine, 38
Methyl cellosolve, 38
Methyl chloride, 38
Methylchloromethylether, 11
Methylcyclopentane, 69
Methylene chloride, 38
Methyl ethyl ketone, 38, 62, 183
Methyl isobutyl ketone, 69
Methyl methacrylate, 69

Met-L-X fire extinguishers, 45–46
Mineral acids, 87, 88, 133
Mine Safety and Health Administration
    (MSHA), 38, 42, 43, 140
Minimization of contamination, 155–156
Minimization of exposure, 14
Mists, 106, 134, *see also* specific types
Mixtures, *see also* specific types
    of gases, 109
    of hazardous wastes, 182
    hypergolic, 65–66
Molecular oxygen, 76
Monitoring, 107–110, 144, *see also* specific
    types
    exposure, *see* Exposure monitoring
Monoethanolamine, 38
Morpholine, 38
MSDSs, *see* Material safety data sheets
MSHA, *see* Mine Safety and Health
    Administration
Multiple emergencies, 134

N

Naphthalene, 38, 60
α-Naphthylamine, 11
β-Naphthylamine, 11
National Fire Protection Association
    (NFPA), 5, 60, 61, 63
    hazardous chemical identification system
    of, 54–58
National Institute for Occupational Safety
    and Health (NIOSH), 1–2
    Certified Equipment List of, 39
    IDLH and, 140
    Pocket Guide to Chemical Hazards, 140, 141
    record keeping and, 198
    respiratory protection and, 38, 39, 42, 43
    Standard Completion Program of, 140
National Toxicology Program (NTP), 23
Nephrotoxins, 9, *see also* specific types
Neurotoxins, 9, *see also* specific types
Neutralization, 154
New-employee training, 192–194
NFPA, *see* National Fire Protection
    Association
NIOSH, *see* National Institute for Occupa-
    tional Safety and Health
Nitrates, 52, 67, 84, 151–152, *see also*
    specific types
Nitric acid, 38, 53, 65, 72, 83, 87, 151
Nitrides, 84, *see also* specific types

Nitriles, 84, *see also* specific types
Nitrites, 84, *see also* specific types
Nitrobenzene, 183
4-Nitrobiphenyl, 11
Nitro compounds, 67, 68, *see also* specific types
Nitrogen, 51
Nitrogen-containing compounds, 68, *see also* specific types
Nitrogen dioxide, 77
Nitrogen tetroxide, 65
Nitroglycerine, 49, 68
Nitromethane, 68
N-Nitrosodimethylamine, 11
Nonflammable gases, 50, 51, 55, *see also* specific types
Nonmetal halides, 71, *see also* specific types
Nonmetal oxides, 71, *see also* specific types
Nonserious violation under OSHAct, 4
Nonspecific source wastes, 181
NTP, *see* National Toxicology Program

O

Occupational history of employees, 118
Occupational Safety and Health Act
    (OSHAct), 1–7, *see also* Occupational Safety and Health Administration (OSHA)
    employer's responsibilities under, 13–18
    objectives of, 2–3
    penalties for violations of, 4–5, 13
    record keeping under, 18, 123, 194, 195–198
    state regulations and, 6–7
    violations of, 4–5, 13
Occupational Safety and Health Administration (OSHA), 1, 14, 32, 139, *see also* Occupational Safety and Health Act (OSHAct)
    chemical hygiene plan under, 15–16, 19–23
    design standards of, 6
    emergency planning and, 132
    exposure monitoring and, *see* Exposure monitoring
    flammable chemical definitions of, 61–63
    Hazard Communication Standard of, 10, 17, 196
    horizontal-type standards of, 6
    inspections of, 3–4
    Laboratory Standard of, *see* Laboratory Standard of OSHA

Maritime Standard of, 139
    medical record keeping under, 123, 195–197
    performance standards of, 6, 9, 92
    personal protective equipment standards of, 32, 46
    record keeping under, 18, 123, 194–198
    respirator standard of, 18, 36
    specification standards of, 92
    Standard Completion Program of, 140
    standards of, 5–6, *see also* specific types
        design, 6
        Hazard Communication, 10, 17, 196
        horizontal-type, 6
        laboratory, *see* Laboratory Standard of OSHA
        Maritime, 139
        performance, 6, 9, 92
        on personal protective equipment, 32, 46
        respirator, 18, 36
        specification, 92
        ventilation, 91, 92
        vertical-type, 6
    training and, 20–21, 189, 194
    ventilation standards of, 91, 92
    vertical-type standards of, 6
Occupational Safety and Health Review Commission (OSHRC), 1
Octacarbonyldicobalt, 65
Off-site surveys, 144
Oil absorbents, 145
Oil filters, 42
Oleum, 49, 154
On-site surveys, 144
On-site waste storage, 185–186
On-the-job training, 190, 193
Open-circuit SCBA, 43, 44
Optical emission spectrophotometry, 115
Organic nitrates, 67, *see also* specific types
Organic peroxides, 50, 67–69, 152, *see also* specific types
Organometallics, 46, 71, *see also* specific types
Organophosphorus compounds, 175, *see also* specific types
ORM, *see* Other regulated material
OSHA, *see* Occupational Safety and Health Administration
OSHAct, *see* Occupational Safety and Health Act
OSHRC, *see* Occupational Safety and Health Review Commission

Other regulated material (ORM), 53, *see also* specific types
Oxidation capability, 58
Oxides, 11, 53, 62, 71, 72, 84, 157, *see also* specific types
Oxidizers, 48–50, 52, 55, 73, *see also* specific types
  defined, 52
  emergency response procedures for, 151–152
  fires caused by, 151
  solid, 151
  storage of, 82, 83, 87, 88
Oxygen, 50, 51, 55, 67, 71, 76, 139
Oxygen deficiency, 43, 140
Oxygen dissociation, 76
Oxygen supply, 76
Ozone, 77

P

Paper caps, 49
Paramagnetic analysis, 139
Parathion, 52
Particulate matter, 106, 109, 110, *see also* specific types
Passive exposure monitoring, 110, 112–114
Patched containers, 148
PELs, *see* Permissible exposure limits
Pentacarbonyliron, 65
Pentachlorophenol, 183
n-Pentane, 62
Peracids, 84, *see also* specific types
Perchlorates, 84, *see also* specific types
Perchloric acid, 38, 65, 73, 84, 87, 151
Perchloric acid fume hoods, 102
Perchloro compounds, 68, *see also* specific types
Performance standards of OSHA, 6, 9, 92
Permanganates, 52, 65, 84, *see also* specific types
Permissible exposure limits (PELs), 14, 17, 18, 20, 21
Peroxides, 68–70, *see also* specific types
  diacyl, 70
  emergency response procedures for, 151, 152
  inorganic, 52
  liquid, 152
  organic, 50, 67–69, 152
  solid, 152
  storage of, 84
Peroxidizable chemicals, 69, *see also* specific types

Personal hygiene rules, 27
Personal protective equipment, 21, 26, 31–46, *see also* specific types
  for body, 34–35
  in emergency response, 44–46, 134, 143
  for eyes, 32–34, 44–45, 73–74
  for face, 32–34
  for hands, 35–38
  inspection of, 31
  maintenance of, 31
  OSHA standards on, 32, 46
  respiratory, 36–44
  safety, 44–46
  selection of, 31
  standards on, 32
  storage of, 31
  training in use of, 31
  use of, 31
Personnel monitoring, 108
Phenols, 38, 49, 84, *see also* specific types
Phosgene, 51, 77
Phosphates, 84, *see also* specific types
Phosphides, 84, *see also* specific types
Phosphoric acid, 38
Phosphorus, 49, 84, 150–151, 175
Phosphorus pentachloride, 71
Phosphorus pentoxide, 72, 84
Phosphoryl chloride, 71
Photoionization detectors, 138
Physical examination, *see also* Medical examination
  of laboratory workers, 118–120
  of victims, 167
Picrates, 68, *see also* specific types
Picric acid, 67, 68, 152
Pipe clamps, 148
Pipe test plugs, 148–149
Pitot traverse of ducts, 99
Placarding system, DOT, 54, 55
Plastic aprons, 35
Plug-N-Dike putty, 147
Pneumoconiosis, 77
Pocket Guide to Chemical Hazards, NIOSH, 140, 141
Poison gases, 50, 51, 54, 55, 152, 153, *see also* specific types
Poisons, 48, 52, 172, *see also* specific types
  emergency response procedures for, 152–153
  gas, 50, 51, 54, 55, 152, 153
  placards for, 55
Polio virus, 52
Polymerizable compounds, 68, *see also* specific types

Polymerization catalysts, 87, *see also* specific types
Potassium, 45, 52, 71, 151
Potassium arsenate, 52
Potassium chlorite, 68
Potassium chromate, 53
Potassium hydroxide, 38
Potassium permangante, 65
Pre-employment medical examination, 118–120
Preservation of records, 196, 198
Pressure, 93, 99
Pressure vessels, 5
Pressurized systems, 29
Prevention, 60, 65, 125, 127
Prior approval, 21
Prismane, 68
Project director, 16
Propane, 51
β-Propiolactone, 11
Propyl acetate, 62
Propylene dichloride, 38
Propylene oxide, 62
Protection factor, 39
Protective equipment, *see* Personal protective equipment
Pulmonary disease, 77
Pyridine, 62, 183
Pyrophoric chemicals, 64–65, *see also* specific types
Pyrotechnic signal devices, 49

Q

Qualitative respirator fit test, 39

R

Radioactive materials, 48, 50, 52–55, *see also* specific types
Radioactivity, 58
Radio communication, 128
Radioisotope fume hoods, 102
RCRA, *see* Resource Conservation and Recovery Act
Reactive chemicals, 66–71, *see also* Reactivity; specific types
water-, *see* Water-reactive chemicals
Reactivity, 57–58, 66, 67, 180–182, *see also* Reactive chemicals
Receiving hoods, 95
Record keeping, 18, 123, 194–198
Reducing agents, 70, 83, *see also* specific types

Reproductive toxins, 9, 23, *see also* specific types
Rescue, 130–131
Resource Conservation and Recovery Act (RCRA), 179–181, 184, 185
Respirators, 18, 36–44, *see also* specific types
air-line (supplied-air), 42
air-purifying, 40–42
emergency, 39
in emergency response, 134, 153
nonemergency, 40
OSHA standards for, 18, 36
qualitative fit test for, 39
selection of, 41
supplied-air, 42, 141
"Type C", 42
Respiratory protection, *see* Respirators
Respiratory tract cancer, 78
Response unit team leader, 144
Restricted (exclusion) zone, 135–136, 159
Ribidium, 71
Risk underestimation, 14
Rotating vane anemometers, 98
Routes of exposure, 75, *see also* specific types
Rubber aprons, 35, 73, 74
Rubber gloves, 36–38

S

Safety awareness, 25
Safety equipment, 44–46, *see also* specific types
Safety fuses, 49
Safety glasses, 33
Safety matches, 51
Safety planning, 144
Safety practices, 25–30, *see also* specific types
Safety rules, 25–29
Safety shields, 26, 33
Safety showers, 44, 45, 74
Safety spectacles, 33
Sampling, 116
analysis in, 115
continuous (integrated), 108
decontamination and, 158–163
efficiency of, 109–110
electronic circuitry in, 138
of gases, 109
grab (instantaneous), 108, 111
instantaneous (grab), 108, 111
integrated (continuous), 111

laboratory analysis in, 138–140
number of samples in, 110
of particulate matter, 109
temperature and, 107
volume in, 108
Sampling devices, 108, 109, 112, *see also* specific types
SARs, *see* Supplied-air respirators
SCBA, *see* Self-contained breathing apparatus
Selenides, 84, *see also* specific types
Selenium, 53, 183
Self-contained breathing apparatus (SCBA), 39, 42, 43
closed-circuit, 43, 44
in emergency response, 134, 135, 141, 159
entry-and-escape, 43–44
escape-only, 43
heat stress and, 172
open-circuit, 43, 44
removal of, 161
Sensitizers, 9, *see also* specific types
Serious violation under OSHAct, 4
Shaving cream, 53
Shipment of hazardous wastes, 186–188
Shock-sensitive chemicals, 67, *see also* specific types
Shoes, 74
Side shields, 33
Silicates, 84, *see also* specific types
Silver, 183
Silver amide, 68
Silver azide, 68
Silver fulminate, 68
Silver nitride, 68
Site characterization, 143–144
Site control, 129
Site maps, 128
Site safety plan, 144
Site security, 129
Skin absorption of chemicals, 75, 141
Skin contamination first aid, 176
Skin protection, 135
Small arms ammunition, 49
Small quantity generators, 184–185
Smoke, 106
Smokeless powders, 49
Smoke tube tracer test, 97–98
Smoking, 26
Soda ash, 151
Sodium, 45, 52, 71, 151
Sodium bisulfite, 70

Sodium carbonate, 157
Sodium hydrosulfide, 51
Sodium hydroxide, 38, 72, 157
Sodium hypochlorite, 38, 157
Sodium nitrate, 151
Solid peroxides, 152
Solid potassium permanganate, 65
Solids, 51–52, 147, 151, *see also* specific types
flammable, *see* Flammable solids
Solubility, 77
"Solution by dilution", 154
"Solution by neutralization", 154
Solvents, 69, 70, *see also* specific types
Special hazards, 58, *see also* specific types
Specification standards of OSHA, 92
Specific source wastes, 181
Spectrophotometry, 115, *see also* specific types
Spectroscopy, 115, 138, *see also* specific types
Spills, 21, 26, 28, 73, 74, *see also* specific types
cleanup of, 182–183
containment of, 145–147
liquid, 145–147
solid material, 147
Spine immobilization, 167
Spontaneously combustible solids, 51, *see also* specific types
Stable (unreactive) chemicals, 66
Standard operating procedures, 19, 155
State regulations, 6–7
Static pressure, 93, 99
Stockrooms, 84–85
Storage, 81–89
of carcinogens, 89
compatibility and, 82–84
of corrosives, 88
of emergency equipment, 130, 133
of equipment, 31, 130, 133
of flammable chemicals, 86–88
of hazardous wastes, 185–186
of highly toxic chemicals, 89
inventory control and, 81–82, 89
of oxidizers, 87, 88
of personal protective equipment, 31
storerooms for, 84–85
of toxic chemicals, 88–89
ventilation for, 88
of wastes, 185–186
of water-reactive chemicals, 87–88
Storerooms, 84–85

Strained ring compounds, 68, *see also* specific types
Strong acids, 72, *see also* specific types
Strong bases, 72, *see also* specific types
Styrene, 49, 62, 68, 69
Substance-specific standards, 10
Suction, 93
Sulfates, 84, *see also* specific types
Sulfides, 84, *see also* specific types
Sulfinyl chloride, 71
Sulfites, 84, *see also* specific types
Sulfoxides, 84, *see also* specific types
Sulfur, 51, 84
Sulfuric acid, 38, 53, 65, 72, 87, 151, 154
Superfund, 188
Supervisors, 15, 16, 25, 130, 191, 193
Supplied-air respirators (SARs), 42, 141
Support zone, 136–137, 159

T

TCLP, *see* Toxicity characteristic leaching procedure
Tear gas candles, 52
Tetrachloroethyleneyridine, 183
Tetraethyl lead, 76
Tetrafluoroethylene, 69
Tetrahydrofuran, 62, 69
Tetrahydronaphthalene, 69
Thiosulfates, 84, *see also* specific types
Threshold limit values (TLVs), 6, 14, 17, 18, 20, 21
   defined, 109
   in emergency response, 141
   in exposure monitoring, 109, 115
   heat stress and, 174
   hoods and, 102
Threshold Limit Values and Biological Exposure Indices, ACGIH, 141, 174
Threshold limit value-ceiling (TLV-C), 109
Threshold limit value-short-term exposure limit (TLV-STEL), 109
Threshold limit value-time-weighted average (TLV-TWA), 109
Titanium, 65
Titanium powder, 64
TLV, *see* Threshold limit values
TLV-C, *see* Threshold limit value-ceiling
TLV-STEL, *see* Threshold limit value-short-term exposure limit
TLV-TWA, *see* Threshold limit value-time-weighted average

Toluene, 38, 62, 70
Toluene diisocyanate, 78
Tool drop, 159–160
Total pressure, 93
Toxaphene, 183
Toxic chemicals, 9, *see also* specific types
   action of, 75–76
   emergency response procedures for, 152–153
   handling procedures for, 74–79
   storage of, 88–89
   water solubility of, 77
Toxicity, 52, 74, 78, 102, 122
Toxicity characteristic leaching procedure (TCLP), 180, 182, 183
Toxicology, 74
Toxic wastes, 180, *see also* Hazardous wastes
Trade secrets, 197
Trainers, 190–191
Training, 20–21, 189–194
   development of program for, 189–190
   for emergencies, 131–132
   in fire extinguisher use, 46
   in hood usage, 103–104
   instructors in, 190–191
   materials for, 190
   methods in, 190
   needs for, 189–190
   new-employee, 192–194
   objectives of, 189–190
   on-the-job, 190, 193
   outlines for, 190
   program development for, 189–190
   in protective equipment usage, 31
   record keeping in, 194
   for supervisors, 191
   trainers for, 190–191
   for transferred employees, 194
Transfer of records, 198
Transferred-employee training, 194
Tremolite, 11
Triage, 167–169
Tributyl phosphine, 64
Trichloroethylene, 38, 183
2,4,5-Trichlorophenol, 183
2,4,6-Trichlorophenol, 183
Tricresyl phosphate, 38
Triethanolamine, 38
Triethylamine, 65
Trimethylaluminum, 64, 65
Trimethylphosphine, 65

Trinitrotoluene, 38, 67, 68
"Type C" respirators, 42

U

UFL, *see* Upper flammable limit
Ultraviolet photoionization detectors, 138
Unattended operations, 29
Underestimation of risk, 14
Uniform hazardous waste manifest, 187–188
United Nations numbering systems, 47–48
Universal absorbents, 145
Unreactive (stable) chemicals, 66
Unstable chemicals, 66–70, *see also* specific
    types
Upper flammable limit (UFL), 60, 61
Uranium hexafluoride, 49
Utility services control, 129

V

Vapors, 106, 110, 112, 134, 138, 154, *see
    also* specific types
Velocity pressure, 93
Velometers, 98
Ventilation, 20, 64, 91–104, *see also*
    specific types
  adequate, 14, 88
  dilution, 91–92
  ducts in, 93, 99
  general (dilution), 91–92
  hoods in, *see* Hoods
  local exhaust and, 91–93
  OSHA standards for, 91, 92
  for storage, 88
Vertical-type standards of OSHA, 6
Victim handling
  assessment in, 166–167
  consciousness level assessment in, 170
  decontamination in, 170–172
  evacuation in, 172
  first aid in, 175–177
  initial stabilization in, 169–171
  stages of, 165–166
  triage in, 167–169

Vinyl acetate, 69
Vinyl acetylene, 69
Vinyl chloride, 11, 49, 68, 69, 78, 183
Vinylidene chloride, 69
Vinyl pyridine, 69
Viruses, 52, *see also* specific types

W

Walsh-Healey Act, 5–6
Wastes, *see* Hazardous wastes
Water fire extinguishers, 45, 150
Water fogs, 150
Water-reactive chemicals, *see also* Water
    reactivity; specific types
  emergency response procedures for, 151
  handling procedures for, 66, 70–71
  solid, 51–52, 151, *see also* specific types
  storage of, 87–88
Water reactivity, 58, *see also* Water-reactive
    chemicals
Water solubility, 77
Wedges, 148
Wet feed, 53
Williams-Steiger Occupational Safety and
    Health Act, *see* Occupational Safety
    and Health Act (OSHA)
Working alone, 29
Work practices, 21

X

X-ray fluorescence, 115
X-rays, 196
Xylene, 62
Xylyl bromide, 52

Z

Zirconium, 65
Zones of hazard, 135–137
  of personal protective equipment, 31
  storerooms for, 84–85
  of toxic chemicals, 88–89